THRIVING
WITH MICROBES

THRIVING
WITH MICROBES

The Unseen Intelligence Within and Around Us

Sputnik Futures

TILLER PRESS

New York   London   Toronto   Sydney   New Delhi

TILLER PRESS

An Imprint of Simon & Schuster, Inc.
1230 Avenue of the Americas
New York, NY 10020

First Tiller Press trade paperback edition October 2021

TILLER PRESS and colophon are registered trademarks of Simon & Schuster, Inc.

For information about special discounts for bulk purchases, please contact Simon & Schuster Special Sales at 1-866-506-1949 or business@simonandschuster.com.

The Simon & Schuster Speakers Bureau can bring authors to your live event. For more information or to book an event, contact the Simon & Schuster Speakers Bureau at 1-866-248-3049 or visit our website at www.simonspeakers.com.

Interior design by Jennifer Chung

Manufactured in Malaysia

10 9 8 7 6 5 4 3 2 1

Library of Congress Cataloging-in-Publication Data
Names: Futures, Sputnik, author.
Title: Thriving with microbes : the unseen intelligence within and around us / Sputnik Futures. Description: New York, NY : Tiller Press, 2021. | Series: Alice in futureland ; book 3 | Includes bibliographical references and index. Identifiers: LCCN 2021005833 (print) | LCCN 2021005834 (ebook) | ISBN 9781982172640 (paperback) | ISBN 9781982172671 (ebook) Subjects: LCSH: Bacteria. | Microorganisms. Classification: LCC QR74.8 .F88 2021 (print) | LCC QR74.8 (ebook) | DDC 579.3—dc23
LC record available at https://lccn.loc.gov/2021005833
LC ebook record available at https://lccn.loc.gov/2021005834

ISBN 978-1-9821-7264-0
ISBN 978-1-9821-7267-1 (ebook)

To the past, present, and future visionaries
who fearlessly explore the beautiful
and the beastly microbes in our world.
And to the microbes that make us "human."

"Hello, *I am Alice, and I am always in a state of wander."*

Alice in Futureland is a book series that asks you to wander into possible, probable, plausible, provocative futures.

Consider this book a guide.

Inside, you will discover extraordinary ideas: a cross-pollination of art, science, and culture. Alice's aim is to give the future a platform for expression, so *everyone* can make sense of it—and help create it.

When speculating about the future, it's easy to get lost in the volume of information. That's where this book comes in. Alice is designed to break the static flow with a dynamic reading experience, where experimentation and exploration meet.

The ultimate purpose of the Alice series is to foster curiosity.

To enliven our present.

To be accessible to everyone.

To allow for exploration.

And to incite optimism.

So, *wheeeeeeeeeee*, down the rabbit hole we go!

CONTENTS

00. INTRODUCTION
Why Bacteria Get a Bad Rap

Microbes rule the world, so much so that saying the word "microbes" is like saying "light"—there are so many different forms and varieties in it, and all are essential to life. The term "microbes" refers to any small organism that is microscopic and generally unseen by the human eye. Microbiology is the study of these microorganisms; some are friendly but seem to have evil twins, and others are still a mystery. Microbes include the beautiful and badass bacteria that live in and on us (and just about everything else on the planet); fantastic fungi that communicate to one another, and that we manipulate for food, medicine, and materials; the (sometimes) yummy yeast we have cultivated for fermented foods; and even the pretentious and often freeloading parasitic protozoa, which we try to avoid. Much like a pop star today, microbes have many followers, admirers, frenemies, and haters.

Microbes are a magical kingdom coexisting with us in every aspect of life. And we are harnessing them to become our factories and engineers; our personalized medicine; both our chefs and our food; our architects and designers; and our partners in fighting climate change and cleaning up the environment. We are inextricably bedded with microbes, you see. We are their spaceships—they are our fellow travelers here on Earth, and we in turn will bring them to the new habitats we jet toward in the cosmos (trip to Mars, anyone?).

Say Hello to Your Microbiome

You and I, and every pet and animal and plant are made of microbes. Tens of trillions of them, and still counting, comprise our very beings and the environments we live in. Yet humanity has had a love/hate relationship with these remarkable invisible organisms. To anti-bac or not has been the question of the last ten years since the microbiome, the colony of bacteria that lives within and on us, became the talk of gut health, immune boosting, mood disorders, energy, and possibly cognitive disorders like ADD and dementia (more on that in chapter 2). Books, nutritionists, and Instagram feeds became flooded with diets, supplements, and lifestyle routines to balance your microbiome and help you hack your way to better health.

And then in the fall of 2019, along came a novel (new) coronavirus, challenging our view of microbes, putting us into sterilization overdrive, masking, and social distancing from one another and the invisible virus. One day we may look back in wonder at how something so small as a virus nearly silenced and stilled our world, and in its aftermath, reset how we socialize in a "new normal." Virtual or remote life became the foundation of the way we worked, celebrated, and learned, and in the wake of the pandemic, we will perhaps weigh the risks of the one sense that helps us feel our world: touch. Many of us will hesitate with any casual, emotional, and physical contact. But we humans are social beings. We crave connection, we like to touch, and we like to be touched. Our emotional ties will get us through the new normal, and the services we seek for our health, wellness, and entertainment will recover with us. And that goes the same for our current relationship with microbes. During social distancing, our microbiome became a closed community as well, and we will need to rewild our microbiome and resocialize with the outside world again to build up the resilience of the microbes we carry.

Some microbes can be our enemies, and bacteria, the largest member of the microbial community, have a bad reputation lately. But they are also family. All life on Earth is said to derive from a com-

mon ancestor: bacteria. Most of life's history is microbial. Bacteria have a remarkable capacity to combine their bodies with other organisms, forming alliances that may or may not be permanent. This concept is called "symbiogenesis" and is a major source of evolutionary change on Earth. Acknowledging that bacteria are our ancestors and are in fact essential to our survival as organisms—for example, without the bacterially derived mitochondria that live in our cells we cannot breathe—proves how interdependent and interconnected life on Earth really is.

From Epidemic to Pandemic

The novel coronavirus that began spreading in Wuhan, China, was an epidemic at first, as it caused a rapid increase in new infections or cases. Once it traveled, or as New York State governor Andrew Cuomo claimed during one of his weekly press meetings, "it got on a plane from China" and spread across several countries, this regional virus became a global pandemic. We don't mean to remind you once again of the fear, bravery, loss, mayhem, disinformation, and upheaval it caused—everything and anything became stalled, reversed. We will almost certainly feel the reverberations of this pandemic for years to come. But despite being catastrophic in many ways—the number of lives lost, the disruption of work, education, and people's livelihoods—the pandemic also taught us how to better understand and respect the role of microbes in our world.

LUCA

Scientists have discovered that we humans have a "Last Universal Common Ancestor," a LUCA in science lingo, that is a single-celled, bacterium-like organism, estimated to have lived tucked away in deep-sea vents here on the Earth around 3.5 billion years ago. Building off a theory that bacteria and archaea (bacteria-like organisms) were the earliest of Earth's inhabitants, evolutionary biologist William F. Martin set out to determine the nature of the *original* organism that both the bacteria and the archaea came from. Together with his team at Heinrich Heine University in Düsseldorf, Germany, he investigated microbes' protein-coding genes, which have been catalogued in DNA databanks over the past twenty years.[2]

Thanks to AI, new decoding machines, and human tenacity, science now has a wealth of gene sequences from thousands of microbes, about six million genes thus far—the holy grail of bacterial origins.[3] And science has only begun to tap this primordial intelligence. Microbes have extraordinary adaptive and altruistic behavior; they communicate through chemical signaling to form bonds, help each other, and even sacrifice themselves for the greater good of their colony. Our knowledge of CRISPR-Cas9 gene editing came from studying the DNA of bacteria and archaea, teaching us how to edit genetic sequences used today in biological research, biotechnology, and the treatment of diseases. Microbes are also teaching us how to ferment our foods, create new materials, and improve our health and that of the environment. Yes, all that and more from mighty, minute, single-cell intelligent life-forms.

Evolution, Creation... or LUCA?

Charles Darwin was the first to propose the "universal common ancestor" theory more than 150 years ago. Today, a large body of scientists have surmised that all species on this planet come from just three groups: bacteria, archaea (bacteria-like microbes), and eukaryotes (the multicellular groups that include plants, animals, and humans). But where did those three groups come from? The answer is in the genomes of all modern organisms. There is no specific fossil evidence of LUCA, so scientists have turned to the genomic data available, identifying a set of 355 genes that are shared among bacteria, archaea, and eukaryotes, suggesting they would have been present in our LUCA. These genes reveal a "complex life-form with many coadapted features, including transcription and translation mechanisms to convert information from DNA to RNA to proteins." The 2016 study by the team at the Institute of Molecular Evolution, Heinrich Heine University, Düsseldorf, Germany, proposed that LUCA probably lived in "high-temperature water of deep-sea vents near ocean-floor magma flows."[4] Another study in 2018 from the University of Bristol that applied a molecular clock model, claims LUCA emerged around 4.5 billion years ago, shortly after the believed formation of the Earth some 4.6 billion years ago.[5]

Okay, so maybe you're still thinking, "How can I vibe with microbes while living through the pandemic or outbreaks like the common flu?" When we began writing this book while still quarantining from COVID-19, there were many rumors and misunderstandings

MY SECRET
PLAN
TO RULE
THE WORLD

about this new strain of coronavirus, with social media at first abuzz with bacteria as the culprit. That has since been debunked (good news for bacteria). There's a simple difference between bacteria and viruses: bacteria are living single-cell organisms, and viruses (smaller than bacteria) are not living—they need a living host in order to multiply and spread. They attack and take over the host cells they latch on to, often hiding away for a free ride. Viruses like COVID-19 are not considered living organisms; however, they are still classified as microorganisms according to the US Department of Health and Human Services.[6] So it's not all microbes' fault—and we aim to show you just how social and beautiful they can be.

Microbes Outnumber the Stars

Trillions of microbes are estimated to make Earth—and our bodies—their home. And that number may be conservative, considering they're microscopic and perhaps the most uncountable life-forms on Earth (and not all of us walk around with a microscope or app to see them—at least not yet). Bacteria, archaea, protists, and fungi—collectively called microbial taxa—are "the most abundant, widespread and longest-evolving forms of life on the planet" according to Jay T. Lennon, a professor of biology at

Indiana University, Bloomington, and Kenneth J. Locey, a faculty member at Diné College in Arizona. In their article for *Aeon*, the authors estimated that the "number of microbial cells on Earth hovers around a nonillion (10^{30}), a number that outstrips imagination and exceeds the estimated number of stars in the Universe."[7]

Who are these micro-things that outnumber our stars? Forgive us as we simplify the science (and decades of microbiology research) for a quick introduction.

There are seven types of microorganisms: bacteria, archaea, protozoa, algae, fungi, viruses, and multicellular animal parasites (helminths). Microbes may exist either in single-celled form or a colony of cells, but keep in mind that single-celled microbes such as protozoa usually latch on to another organism to thrive—microbes don't typically operate alone. Each has a certain type of character, from its cellular composition, to how it moves, morphs, and reproduces.[8] For this wondrous look at the mischievous and magnanimous microbes, we will concentrate on just a few types: bacteria, algae, fungi—and yes, a bit on viruses as well.

If Bacteria Had an Instagram Feed, They Would Break the Internet

For decades we have feared bacteria, given them a bad rap, called them "germs," yet there are way more of them than us—or anything else on this planet. The number of bacteria on earth is estimated to be 5,000,000,000,000,000,000,000,000,000,000. This is five million trillion trillion, or 5 × 10 to the 30th power.[9]

Bacteria are the most social of the microbial world, but they don't often win popularity contests. Bacteria can live in a wide variety of environments, including inside our bodies, where they help us with our biological processes like digesting food. Bacteria often encourage other friendlies to join their tribe and work together for the greater good. But like any social clique, some types of bacteria can be toxic, bad-mouthing each other and recruiting outcasts to soldier up for an invasion. The havoc these nasty tribes wreak include serious diseases like pneumonia (culprit: *Streptococcus pneumoniae*), meningitis (violator: *Haemophilus influenzae*), strep throat (perpetrator: *Group A Streptococcus*), food poisoning (invader: *Escherichia coli* and *Salmonella*), and other infections.[10] But here's the thing: both good and bad bacteria coexist, and it's when the bad bacteria start rallying their peeps to outnumber the good that causes havoc, and can become a menace as prolific as COVID-19.

You Are a Superorganism

So you think you're human? Made of blood, cells, flesh, thoughts, and emotions? Think again. Humans are superorganisms, with literally trillions of microbes taking up residence inside your body, on your skin, and in your hair. Our bodies are home to an invisible universe called the microbiome. Several scientists posit that we are only 10 percent human and 90 percent bacteria! Even if the ratio isn't quite that extreme, it's true that bacteria well outnumber your body's own cells. The Spanish Society for Biochemistry and Molecular Biology estimates that more than ten thousand species of bacteria, of which more than 99 percent are non-pathogenic, inhabit a healthy human body.[11] To put that into perspective, it's been estimated that if we were to line up the bacteria in a human body it could loop the Earth two and a half times.[12] (Makes you wonder, are we really humankind or bacteriakind?)

The late biologist Lynn Margulis posited that species arise from symbiogenesis—that evolution is social—based on her study of the bacterial world. She is credited for the "endosymbiont theory," which asserts that the mitochondria in our bodies (the organelles in our cells that help energy production and respiration) and the chloroplasts of plants evolved from microbes. Margulis's research showed evidence that mitochondria evolved from aerobic bacteria called Proteobacteria, and chloroplasts evolved from endosymbiotic cyanobacteria.[13] It seems that cooperative, friendly, energy-sharing bacteria have helped our cells and plant life around us breathe and reproduce. Yes, we breathe because of bacteria.

The concept of the human microbiome is not that old; it was first suggested by Joshua Lederberg in 2001 "to signify the ecological community of commensal, symbiotic, and pathogenic microorganisms that literally share our body space."[14] Like the citizen scientists who helped fast-track the sequencing of the human genome with their DNA samples, we owe much to everyone who had the interest and the gumption to provide specimens for microbiome research. In just over a decade, research in the human microbiome in the United States alone has surpassed $1.7 billion.[15] Science has gone from identifying the "garden" of bacteria in our bodies to recognizing that our gut microbiota can influence the function of our brain, our immune system, and even our moods. In fact, we may "love" because of a probiotic microbe found in human breast milk that elevates oxytocin, our feel-good hormone.[16] (We are hoping by now you are feeling some love for your microbes!)

Sampling for your microbiome is more complex than just a buc-cal swap you do for DNA testing— a gut microbiome test requires a fecal sample, not a pleasant process for many. But the results from gut microbial tests, as well as those of other body parts, have helped expand our understanding of the diversity of microbes that support or hinder our overall health. There are significant advantages to knowing the different microbial colonies within your body, from your brain and mouth to your gut and private parts. Within the next decade we will undoubtedly be seeing more personalized therapeutics and drugs thanks to a highly intelligent triage: your microbiome!

Humans aren't alone in their microbiomes either. Your pet has their own microbial colonies. Your home has its own microscopic neighborhoods on every surface. As a matter of fact, everything— and yes, we mean *everything*–has a biome of sorts. Your car, the subway, a plane, the buildings and stores you walk in and out of, whether in the city or the suburbs. And those various "biomes" influence your microbiome and, by extension, your health. But how these biomes impact your health goes well beyond just individual

Did You Know?

· Bacteria are called "prokaryotes," because they're single-celled organisms that carry a small amount of genetic material in the form of a single molecule, or chromosome, of circular DNA.[17]

· Bacteria found inside all animals combined (including humans) make up less than 1 percent of the 5 million trillion trillion bacteria in the world.

· The greatest numbers of bacteria live in the subsurface, soil, and oceans of Earth. They have been found forty miles high in the atmosphere and miles beneath the ocean floor.

· Researchers estimate that the total amount of bacterial carbon in the soil and subsurface would be equal to a staggering 5×10^{17} g, or the weight of the United Kingdom.[18]

bacteria. It is your total lifestyle exposure, or what scientists are calling the exposome: the totality of your genome, microbiome, diet, lifestyle, and environmental impact on your health and wellness.

Microbes to the Rescue!

The more we learn about the microbiome, the clearer it becomes that it is *the* gateway to health. Traditionally, medicine has focused on eradicating the bacteria that make us sick (hello, antibiotics!), but the overuse of things like antibiotics and sanitizers that claim to kill 99.9 percent of germs may be creating an antimicrobial resistance (AMR) in humans, impeding the immune fighting powers of our microbes to help fight infection.[19] Scientists suggest that one day your microbiome—whether it's your oral, gut, lung, or skin microbiome—will be a measure doctors will use to give you more precise medication and treatments. These health solutions may include a mixture of "biotics"— prebiotics, probiotics, or postbiotics (and there may be more of these we haven't discovered yet), delivered either by supplement, real foods, or topical solutions. For more serious issues, microbiome "transplants" have shown enormous promise, and there have been successful trials using fecal implants to treat severe cases of colitis, a gastrointestinal disorder.[20] Fecal transplantation (or bacteriotherapy) is the transfer of stool from a healthy donor into the gastrointestinal tract of someone with disease, and, according to Johns Hopkins Medicine, this may be the remedy for our overuse of antibiotics, which kill off too many "good" bacteria in the digestive tract.[21] We're just beginning to discover the healing power of good bacteria for all species, and the inevitability that microbes will be our doctors, surgeons, and medicines.

But wait, what about the highly anti-microbe reality of COVID-19? Well, it's becoming clear that your shrinking social world has also shrunk the social circles of your microbes. It seems the microbial communities in us and the family or roommates we live with are adjusting, and one telltale sign of this may be your bodily smells. Rob Dunn, a biologist at North Carolina State University, explains that if you've been home for some length of time with just a handful of people, you're then just swapping microbes with them and could start to *smell* like your companions, and vice versa. (Sniff test, anyone?)[22]

The Wild Ones Need Names

A number of microbiologists from around the world have come to a consensus that the standard rules for assigning scientific names to bacteria and archaea are due for an upgrade. The current nomenclature of "bacteria" and "archaea"—two of the three large domains of life on Earth—are too generally identified, without regard to the diversity and personality of each individual organism or the collective colony.

So why is this important? In a statement published in *Nature Microbiology* in June 2020, the microbiologists argue that the current code only recognizes those grown from cultures in labs, and not the samples gathered and identified in the wild. The authors explain that "since the 1980s, microbiologists have used genetic sequencing techniques to sample and study DNA of wild microorganisms directly from the environment from habitats such as ocean volcanoes, underground mines, and human skin. Most of these cannot be cultivated in a laboratory, so according to the [International Code of Nomenclature of Prokaryotes] they cannot be officially named." When most of these wild microorganisms are able to be cultured and finally named, imagine what the microbe-naming system will be when you are looking to name, say, a million or more species!

—University of Massachusetts Amherst[23]

Back when we perhaps commuted to work, sat in real-life workshops or concerts, jammed with a CrossFit class, or shopped at the mall, we were in contact with a multitude of diverse microbes. But these days most of us Zoom our way through work or school, rarely venturing into the world beyond a circle of necessary stops. Bottom line: our microbial diversity has changed, and the same thing is happening in the urban and ecological wild. We don't yet know what consequences this will have, but it is inevitable that we need to rebuild our microbiome back to its strong, diverse, and hardworking ecosystem to help fight off all types of infections, viruses, and the common cold.

We Need to Get Micro-Diversified

There's been much discussion and increased awareness around the loss of ecological biodiversity due to our unsustainable habits and the threat of climate change. Biodiversity of species and insects, like the more than twenty thousand species of bees on the planet, of which 80 percent are responsible for helping to produce ninety different food crops (or about one-third of the global food supply chain),[24] is critical for our ecosystems, as well as the biodiversity of soils—not the monoculture that is challenging farming. But at the root of this bio-loss is another larger ecological crisis that we cannot yet visibly detect: the loss of roughly one-third of microbial diversity on Earth.

The long-term impact of this is still not comprehensible, but without drastic changes, such as those proposed by ecologists and environmental advocates, we could be heading toward both an ecological and a microbial collapse. Remember, most species (including ours!) are made of microbial diversity. This possible collapse doesn't bode well for anyone.

But we at Alice in Futureland believe in optimistic futures, the ingenuity of science to stop threats, the biotopians to help solve problems, and, yes, we also believe in the diligence and resilience of hardworking microbes. We have every reason to. Bacteria have been the common "tool" in some of the most advanced discoveries in biotech. From growing high-quality proteins to harvesting energy, pulling excess carbon dioxide (CO_2) from the atmosphere, "eating" plastics to reduce waste, or reversing the impact of antibiotic resistance, bacteria may just one day be our savior.

Given the right circumstances, the biosciences could produce the next wave of innovation in products and services across a range of economic activities in food, agtech, home and personal care, and digital health. Synthetic biology and the perfecting power of clustered regularly interspaced short palindromic repeats (CRISPR) gene editing allows for the creation of entirely new species of organisms. Until now, nature has been the exclusive arbiter of life, death, and evolution. With synthetic biology (also known as biotechnology or intentional biology), we now have the potential to write our own biological future.

Synthetic biology has spawned the bioeconomy, with bioengineers and biodesigners mastering microbes and biological processes to bring us ever closer to an abundant future reality. For example, new foods, fashion materials, beauty and personal care products will be sourced from organism-engineered foundries using fermentation of microbes as its technology.

Today, microbes are our microfactories, micro-engineers, and micro-designers. Pioneers in biodesign are working with bacteria, yeast, and algae to *grow* new buildings, new materials, electricity, medicines, and more. We are entering a biotopian future in which everything will be alive, growing, evolving, changing, and healthy—for us, our earthly and soon interplanetary homes.

Hello Fellow Travelers

It appears that bacteria didn't originate here on Earth—or, at least, so the theory goes. Some five thousand kilometers below the surface of the South Pacific, embedded deep within the ocean floor, scientists have discovered a community of ancient microbes that they believe have lived there for 100 million years. What's more miraculous is when the researchers fed them oxygen back in a lab on land, the microbes loved the oxygen and began to multiply, which made them wonder how they could have survived in an oxygen-deprived seabed. The team of Japanese and American scientists pushed the question further: If ancient life can persist in the deadest parts of the Earth, could it perhaps survive on other planets too? Steven D'Hondt, a professor at the University of Rhode Island who was a part of the expedition team, posits that microbial communities established hundreds of millions of years ago, or even longer, could feasibly persist on Mars (a mere 135.81 million miles away) or Europa (which we know has water vapor, which could mean possible life).[25]

There recently have been several theories around origins of life both terrestrial and extraterrestrial. The January 28, 2020, issue of Live Science reported on two Harvard astrophysicists who proposed a "wild theory of how life might have spread through the universe." How? Extremely hardy microbes hitched a ride with an exploding comet that got ejected out of our solar system billions of years ago, and landed here on Earth—and that may not have been the only time microbe-carrying comet vehicles landed in Earth's early history.[26] Their hypothesis is a take on the theory of panspermia, which dates back to the early 1970s and suggests that life throughout the galaxy and Earth was seeded by space-traveling microbes.

Of course, in our own space explorations, whether manned or robotic, we have probably been seeding the moon and now Mars with microbes from Earth. An inadvertent, interstellar, inter-species cross-pollination of sorts. While NASA and most planetary exploration missions have adhered to sterilization standards that include removal of bacteria on the spacecraft so as not to contaminate equipment or studies, there is still the fact that we humans are the real spaceship: there's no way to prevent us from carrying our trillions of microbes with us. There is much debate on what will happen when we send our first Mars explorers to colonize the Red Planet,

traveling with life support and energy supply systems, food, tools, 3D printers, and more to set up camp. And then there's the basic human function and habit of producing waste, drinking and releasing water . . . you get the picture. We humans may very well be the contaminating virus on Mars. And yet, we will need to bring our own earthly microbial communities if we want to ensure we can maintain our human life and survive as Martians.

What's Out There Is in Here: You Are Conscious Because of Microbes

To be, or not to be . . . one with a microbe? Don't think you have a choice, as you have an ancient virus in your brain that is believed to be the root of conscious thought. According to papers published in the January 2018 issue of *Cell*, a very, very long time ago a virus embed-

ded its genetic code into the genome of four-limbed animals. That snippet of code is still in our brains today, transporting packages of genetic information from nerve cells to their neighbor cells in what looks like carbon copies of the virus itself. (It's all due to the ARC gene. More on that in chapter 6.) The theory is that these packages of information are what fires up our neurons, causing nerves to com-municate over time, akin to the process we recognize as higher-order thinking.

So you see, dear humans, we think, we live, we breathe, we procreate, and we thrive all be-cause of the generous, altruistic (more on that in chapter 1), and mighty microbes.

No matter where we travel, earthly, otherworldly, or in our dreams, we are always, *always* with our microbial family.

Come Together Now

Microbe Tribes and the Holobiont

Renegades, activists, and peacemakers—microbes like to find their squad, cling to the "mother" influencer, create fandoms, and cross talk faster than a Twitterstorm.

Microbes—of which bacteria are the most popular—are far smarter and more resilient than you might think. They have used their influential powers for centuries, and we humans have been in a love-hate relationship with mi-crobes since at least way back in 430 BCE, when the first recorded pandemic swept through Athens during the Peloponnesian War, killing about one-third of the Greek population, including the general Pericles.[1] Epidemiolo-

gists suspect that disease was typhoid fever, which likely traveled from Libya, through Ethiopia and Egypt before making its way to Greece. Today, we know that typhoid fever is caused by the bad-boy Salmonella bacteria that you can ingest from contaminated food or water. Between 1346 and 1353, the Black Death, also known as the Bubonic Plague, was said to have wiped out half of Europe's population, the doing of a strain of the bacterium *Yersinia pestis* that was spread by fleas on infected rodents.[2] The next great wrecking ball of a microbe was the 1918 influenza virus, known as the "Deadliest Flu," often compared to COVID-19 in that it was a "novel," highly infectious virus that spread through respiratory droplets. Medical and scientific experts have been studying the 1918 virus for several years, looking for clues to help fight modern viruses, especially SARS, to which COVID-19 is closely related.[3]

Since then, modern outbreaks like the flu and COVID-19 have forced us to anti-bac up and sterilize everything. But in the last decade, we have just begun to embrace the trillions of microbes that live on and within us, a bacterial colony called the microbiome. These microbes aren't all naughty—actually, some scientists want us to balance, feed, and cultivate our personal microbial tribes. We are constantly immersed in a myriad of microbes, walking through others' microbial clouds, and whether we like it or not, it's time to come together right now.

Oh Behave!

Microbes act like a community of influencers that gain massive followers and sometimes bully their legions to do harm. Their power lies in the gift of gab. Bacteria in particular are great talkers, and their persuasive rhetoric can sway their allies and their frenemies into action. For example, some bacteria in a social situation synchronize their behaviors so that all the members act like one big organism, almost like the world's most popular tweet gains influence through millions and millions of retweets.

Another type of bacterial social behavior promotes more individuality among its group, encouraging diversity that can help with the health of the community or, sometimes, make it unwell. Bacterial communication has been fascinating researchers for years, especially in their attempt to understand how microbes spread, how antibodies can be conceived, and how various environments, from our bodies to the soil to the surfaces of our things, can impact our health and immunity.

Ultimately, it all comes down to the conversations and signaling power of bacteria. Researchers Bonnie L. Bassler and Richard Losick have identified four key methods, "long- and short-range chemical signaling channels; one-way, two-way, and multi-way communication; contact-mediated and contact-inhibited signaling; and the use and spread of misinformation or, more dramatically, even deadly information."[4] We are still learning about the sway bacteria have, and the secrets they share.

Bacteria Cross Talk

When humans coined the term "consensus," bacteria had been practicing it for millennia. Bacteria are the very first operating cooperative on Earth (and perhaps the universe, but more on that in chapter 6). Every member in their microbial cooperative is a worker and stakeholder, vested in protecting the greater good. To that end, bacteria engage in public and private conversations using their own process called quorum sens-ing. It's a cell-to-cell process where bacteria share information about their cell or host to initiate a collective response. Quorum sensing is a motivating rally—it enables bacteria to express certain processes as a team when they're certain that something will have an impact on the environment they're in or the host they're hanging with.

Part of the duties of a bacterium is to detect and elicit a response to autoinducers (AIs), an extra-

cellular form of signaling molecules that act like another form of AI (artificial intelligence). Bacteria's autoinducer AI helps keep tabs on inventory, and changes in the density of population that accumulate in and around their co-op environment. When the bacterial population increases in a co-op, the autoinducers increase as well. This helps the colony monitor cell numbers, and also helps them collectively alter their gene expression—in other words, the more autoinducers, the better they can direct and perform activities that are beneficial to the group. Autoinducers are the chemical signals needed for quorum sensing—bacteria's own Wi-Fi that enables them to detect, signal, connect, and band together for a collective response. There are so many events in our world, both notable and invisible, that are powered by the quorum sensing of bacteria, from the bioluminescent bacteria that make oceans glow, to the bacteria that spread the common cold, to that end-of-day film on your teeth, or the spoiling of your foods.[5, 6]

Interestingly, there's also a social structure in the way bacteria quorum sense and regulate their gene expressions. Recent experiments have shown that in a population of genetically identical microbes—think for example, a family—some individuals in the family may produce autoinducers and others may not. And like a family, different individuals may perform different roles or duties for the unit, some even reserving or decreasing their fitness in order to call on that reserve for increased fitness in stressful conditions, a practice known as biological bet-hedging. The act of bet-hedging is believed to help the overall population to survive when exposed to threats.[7]

Bacteria cohabitate like most other species, living in diverse communities among other cells that may differ from each other, even though they are within a common species. And, just like humans, bacteria have different languages. Researchers have found that each bacteria type has a slightly different autoinducer, like talking on different radio frequencies so their messages don't get mixed up or intercepted.[8]

Bacteria are always collecting information about their surroundings and the microbial citizens living nearby, so they're extremely adept at sensing how many bacteria are in the area. But it's not just the sheer number of "others" that bacteria can detect; it is the proportion of the diverse groups within their environment that bacteria keenly surmise. This is the discovery of a research team at the Max Planck Institute for Terrestrial Microbiology in Marburg and Heidelberg University, a process they identified as "microcensus" in bacteria, where bacteria can suss out a crowd and determine the size of different groups in a mixed population. It is a strategic power bacteria practice to help them make decisions. Ilka Bischofs-Pfeifer, the principal investigator of the research team at the Max Planck Institute, used the analogy of finding a dance partner in a crowded ballroom. She explains that the "sheer number is only of limited relevance to you; it is the gender ratio that tells you how hard it will be to find a dancing partner." Bacteria have the gift of collecting information and detecting that ratio, and that "information about group ratios could help them make decisions and adapt in the best possible way."[9] In other words, find the right dance partners to tango with.

Bacteria Diversify Their Portfolio

This extraordinary gift of sensing the various groups in a mixed crowd is based on molecular signaling systems (autoinducers). The Max Planck Institute research team looked into the signaling in the bacterium *Bacillus subtilis* (a bacterium mostly found in soil and vegetation) known for splitting its population into subgroups, or tribes, that have different properties and duties, diversifying the phenotypes of the whole group like a stockbroker diversifying a portfolio. What the researchers found was that

Bacteria Hedge Their Bets

Scientists have identified that bacteria use a strategy like financial traders use: bet-hedging. You may have heard about hedge funds, which are used as an investment position to offset potential losses or gains that may be incurred by an investment. They can come in many financial forms such as stocks, insurance, options, or futures contracts. The same risk positioning behavior is seen across many organisms, including prokaryotes (the kingdom bacteria belong to). Biological bet-hedging is a mode of risk management for environmental change.[10] It reflects the adaptation organisms practice to allow them to survive in fluctuating environmental conditions and it has a distinct evolutionary advantage. Bacteria are masters of this adaptive behavior—they are known to decrease their fitness levels by reserving their energy and not expending themselves during a "good year" or time of good conditions, in exchange for increasing their endurance during stressful conditions (like a "bad year").[11] It's a strategy that utilizes diversification of bacterial phenotypes within a population in a move to reduce the overall risk of death of all the cells in its population.

Bacillus subtilis produce identically constructed signals, and although the signals were produced by just one subgroup of cells, they were taken up by all bacteria in the *Bacillus subtilis* group. This diversification has an advantage: the variety in the composition of the colony's portfolio allows the bacterium to remain resilient to environmental changes, absorb losses, and even increase its numbers.[12] Bacterial cells have to make a tradeoff between longevity and reproductivity. It is either grow slower but have more ability to live longer, or grow faster, making the bacterial cells more fragile. For a bacteria colony to survive and expand, it employs a division of labor among its constituents, and diversifies its portfolio to invest in both strong producers for the long haul, and the fast growers for rapid expansion.[13]

We Can Talk to Bacteria, and Bacteria Can Talk to Us!

While we are still learning the intricacies of quorum sensing, scientists have made progress in synthetically producing the compounds bacteria use to communicate. It's better than cracking the code—we're not just listening in on their conversations, we can talk to them and, even more exciting, they can talk back.

Why is this important to us? By synthetically producing these compounds, we are joining bacterial conversations, and triggering new conversations that we can control. We are on the verge of using bacteria to work as our personalized medicine. Our skin is the largest organ of the body and hosts microorganisms known as skin flora or skin microbiota. Many of these microbes are bacteria, and around one thousand species of microbes found on the human skin come from nineteen classes of bacteria.[14] Most them are found nesting in the superficial layers of the epidermis and the upper parts of hair follicles. Collectively, this colony is called the skin microbiome. We'll explore more of the wonders of our skin's microbiome in chapter 2, but rest assured, most skin flora are usually nonpathogenic and are either commensal, coexisting neutrally and neither benefiting nor doing us harm, or mutualistic, offering a benefit to us. These commensal and mutualistic bacteria on the skin help prevent transient bad pathogenic organisms from colonizing the skin surface, and they do this by getting their competitive game on. Bacteria have an arsenal of ways they fight off aggressive invaders, either by competing for nutrients, or secreting chemicals against them.[15] Actually, a compound that certain skin-dwelling microbes secrete has been shown to disrupt DNA formation in cancer cells, potentially stopping their growth—perhaps one day we may manipulate our own skin microbes to prevent skin cancer.[16]

The idea of using bacteria to work for us is not a new one. Bacteria in our bodies help us break down food and turn it into

Bacteria Practice (Social) Policing

To maintain their social order, quorum-sensing bacteria deploy a strategy that keeps "cheaters" in line to maintain cooperation in an increasingly hostile environment. Biologists at Indiana University Bloomington, have learned that naturally cooperative bacteria may evolve to behave in ways that block the increase of selfish members of the colony and impede their bids for dominance. Cancer cells are a good example: in humans—the ultimate complex multicellular organism—cancer cells are the "cheaters" who, rather than cooperating, attempt to multiply at the expense of the larger organism. That increase needs to be suppressed by our healthy cells (or sometimes with the support of medical intervention) in order to avoid being taken over or becoming disease. The same goes for bacteria, who need to socially police their good and bad growth to avoid losing their hold in an environment.
—ScienceDaily[17]

nutrients and aid our frontline defenses against infection. We use bacteria to help create vital antibiotics. They have been our chefs, fermenting foods like cheese and yogurt; and we use their fermentation techniques to create compost from our organic waste. Soil bacteria keep our farmlands, fields, and forests nutrient-rich. But it has only been recently that biologists have been able to drop in on the conversation bacteria are having, and manipulate the way bacteria talk to each other, giving us the chance to not only join in but direct the discussion to our benefit.

I'm Listening!

One team of bioengineers at the University of California, San Diego, has done this with the harmless kind of *E. coli* bacteria, reprogramming their conversation process to prevent communication unless one specific external molecule—in this case, the bacterial gene lysis—is present in the population. The researchers found that the lysis gene turns on the chemical signals that trigger bacteria's cross chatter.[18] The beauty of this work is that we now have a new genetically engineered "circuit" that could be used like an app for synthetic biologists to instruct the bacteria they are engineering to deliver drugs or perform other tasks, ultimately putting bacteria to work for us just by flipping a switch to turn on their conversations. According to Arianna Miano, a UC San Diego bioengineering PhD student and lead author of the research paper, "We have just scratched the surface of the potential of this communication system. We are excited to see the applications that will follow by coupling it to the expression of different genes."[19]

Of course, we aren't the only ones eavesdropping on bacteria. In 2018 a team of Princeton researchers discovered that a virus, VP882, can listen in on bacterial conversations and, even more important, twist the discussion to go on the attack. The researchers were able to reengineer the virus to respond to external sensory inputs they provided, rather than the communication molecule that bacteria use, to make the virus kill on demand. The newly programmed VP882 was able to confront bacteria like *E. coli* and diseases such as cholera, and their success suggests that reengineered viruses may help to counteract antibiotic-resistant bacteria.[20]

Our very existence is codependent on microbes, and tapping into their conversations will help us to steer their discussions to help deliver targeted drugs, produce new foods, materials, and more, all of which we will cover in the chapters to come. But first let's look at the intelligence these microorganisms possess, and how we can co-opt bacteria's smart ways.

If you look at the bacterial kingdom as a whole—what a wonderful thing it is. In a sense, it's a huge computer because these guys can share back and forth DNA in a much more promiscuous way than sexually oriented large creatures can.

—Vernor Vinge, retired San Diego State University professor of mathematics, computer scientist, and science fiction author, Sputnik Futures interview, 2000

The Tribes

They may be single-celled microorganisms that don't have a nervous system, or a heart, or a brain, but bacteria are known to exhibit some impressive intelligent behavior. These miniscule life-forms are "Big Thinkers," using a form of intelligent processing. Bacteria use receptors of touch or sense the chemicals released by other bacteria, and they "interpret" that information and then signal others—basically sensing, processing, and then persuading. Bacteria have also been observed to solve problems such as food scarcity; encourage other bacteria to join them and build "cities"; and are amazing survivors often overtaking a host or another colony. Let's dive into the characteristics bacteria display to better understand their symbiotic processing power—and the societies they build, with and without us.

The Big Thinkers

Bacteria are perhaps the most intelligent and persuasive microorganism on the planet. We have learned that through quorum sensing, bacteria can induce behavioral changes in other cells within their population or a mixed-species population. This behavior is often referred to as "microbial intelligence." While we know that bacteria communicate through the release of chemicals, it may help to understand the way this helps them "think." Bacteria cells contain various receptors, and each receptor affects a specific behavior in the bacteria—for example, where to move or whether to become virulent. Genetic sequencing has enabled us to understand how many receptors bacteria have, and the ways individual bacterium use these to sense its surroundings.[21]

Bacteriologists are still investigating how this process of "thinking" originates, by observing how

single-cell bacteria can sense, respond, and adapt as it interacts in a local environment. Findings so far indicate a system of first sensing and then communicating to other bacteria, described as a model of "bacterial neural networks," because the overall communication system of a colony can look like the neural network of the human brain. The human brain makes its connections through a pattern of electrochemical signals that pass from one neuron to another, certain receptors selectively picking up signals, thereby forming a neural network. Bacteria also make connections, sense, and communicate through chemical signals. Like humans, bacteria take inputs from their

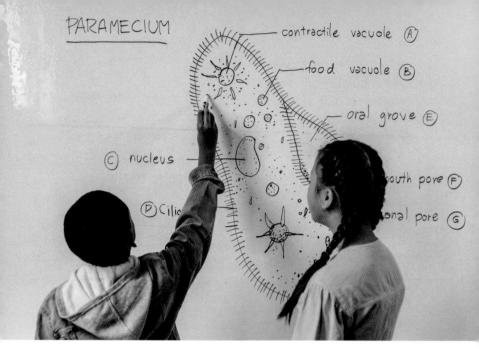

PARAMECIUM

- contractile vacuole (A)
- food vacuole (B)
- oral grove (E)
- mouth pore (F)
- anal pore (G)
- (C) nucleus
- (D) Cilia

environment—for example, sensing the surface they are on or the presence of food. They "process" that input and send out chemical signals that are selectively picked up by other bacteria in their colony. One study demonstrated that individual *Bacillus subtilsi* growing in an area with very little food released their chemical signals warning that "hey, there isn't enough food here for all of us" to other bacteria in the colony. Bacteria who responded to the chemical signaling moved themselves farther away from the other bacteria still in the food-poor area, changing the shape and dynamics of the colony.[22] Researchers at Tel Aviv University in Israel have developed a fractal model of a bacterial colony, going so far as identifying linguistic and social patterns in a colony life cycle.[23]

Bacteria may not have a brain, but they do have memory—and a pretty good one according to biologists from the University of San Diego, California. This microbial memory was discovered in a collective of bacteria, known as "biofilms." The researchers stimulated individual bacterial cells within the biofilm with light when the bacteria were communicating through ion channels produced by the potassium in the chemical signals they release. These ion channels act like phone lines between the bacteria in the biofilm, a very low-level electrical charge similar to how the neurons in our brain work.

The researchers exposed the *Bacillus subtilis* biofilms to light during their communications, thereby encoding a light stimulus during the time of a "task." The light induced changes to the biofilm's ion channels, and consequently transformed the membrane, or overall community of biofilm. But what is most interesting is that hours after the first stimulation, the researchers repeated the light-induced stimulus, and discovered that the bacteria responded. According to Professor Gürol Süel, head of the laboratory at University of California, San Diego, when they disturbed these bacteria with light, the bacteria "remembered and responded differently from that point on." The bacterial cells "remembered" the light exposure and performed their collective activity. While it sounds somewhat simple, the findings show some fascinating parallels between single-cell

Total Recall?

Your average bacterium may have a memory based on its behavior. Evidence is mounting that regulatory networks within bacteria are capable of complex behaviors that have been linked to "memory." Bacteria that have experienced various environmental histories may respond differently to current conditions. These "memory" effects may be regulated by the cell to turn on the memory of fitness to the organism in the evolutionary game it participates in. Researchers propose that history-dependent behavior is a potentially important manifestation of memory, and in the case of bacteria, it presents a framework for quantifying memory in cellular behaviors—and perhaps unveiling new questions about cellular regulation and evolutionary strategy.[24]

Receptors May Determine Bacterial "IQ"

Like some people you may know, certain species of bacteria are known to be, well, "kind of dumb," according to Gladys Alexandre, an associate professor of Biochemistry & Cellular and Molecular Biology at University of Tennessee, Knoxville. Alexandre is referring to the common bacteria *Escherichia coli* (aka *E. coli*), which biologists typically look at when studying sensing and information processing. But she and her team wanted something smarter, a more complex thinker, and they found it in soil bacterium *Azospirillum brasilense*. The intelligence gap between the more common *E. coli* and the special *Azospirillum brasilense* is a matter of how many receptors each individual bacterium has to direct its decision-making process on movement: *E. coli*, five; *Azospirillum brasilense*, forty-eight. As you guessed, the *Azospirillum brasilense* are much "smarter," can detect changes in their environment, and can make more complex decisions about where to move.

—ScienceDaily[25]

organisms and the complex neurons that process memory in the human brain. The researchers are hopeful that this ability to encode memory in bacterial communities may be the premise for the future design of biological computation—basically computing systems with bacteria as the living processor.[26]

A-Mazing Bacteria!

The Problem Solvers

That influential ability of quorum sensing can do more than help individual bacterium detect what other like minds are around them. When these similar bacteria realize they are sisters (or cousins, or friends, or colleagues), they collectively trigger a behavior change in each other. One example of this is in disease-causing bacteria who practice their quorum-sensing prowess to decide if they have the critical mass to launch an attack on their host. And if we can jam their communication signals and prevent them from rallying the troops, it may be possible to neutralize the bacterial mob, something researchers are still exploring.

The idea that microbes have ways of bonding and working together in

The "Glue" That Helps Bacteria Build Cities

The biofilms bacteria build are analogous to urbanization and how we build our cities, as was recently proven by researchers from University of Pennsylvania's School of Dental Medicine who took a "satellite view" following hundreds of bacteria distributed on a tooth enamel–like surface, and watched them grow from individual "settlers" to small bacteria villages. Once the villages grew and their borders met, they joined to make "cities" and some of these cities then merged into "megacities." The researchers had expected that as the different microcolonies met, there might be competition between them, causing the two edges to repel and move apart. But they didn't. What held them together? A glue-like secretion known as extracellular polymeric substances (EPS) that enabled bacteria to pack together closely in the biofilm.

—PhysOrg[27]

stressful environments caught the attention of researchers at Princeton University—actually, as an experiment that began as a dare to design a maze that a physicist at Princeton could not find his way out of. The challenger? Princeton University physicist Robert Austin. Challenge accepted by his grad student Trung Phan. The aha moment: watching Austin's traditional maze-solving strategy—when he hit a dead end, he traced his path back. Which led Phan to question: what if there were a maze with no dead ends? What would Professor Austin do? (Bacteria will be part of this story soon, we promise.) So Phan had a blueprint developed using infinite loops of winding pathways, which a colleague etched on a small silicon chip. Now the player of this new challenge wasn't Austin—it wasn't even human. It was ten *E. coli* bacteria that the researchers trapped in the center of the chip.

Austin's first round was for learning. In this round, their intention was to look at how organisms solve problems—in this case, the fastest route out of a maze. To motivate the ten bacteria, the researchers flooded the chip with bacteria's favorite food, a broth that, according to Phan, "smells like chicken soup." And then the spectators (researchers) watched under the microscope. The bacteria succeeded at the task, basically by eating their way around the maze and reproducing like crazy, so that by end, that original ten grew to become a million bacteria. As the maze-playing bacteria cleared paths of food, Team *E. coli* moved toward the unexplored, broth-rich areas, eventually leading them to successfully evacuate the maze. And it only took about 1 percent of the multiple generations of bacteria to collectively solve the puzzle in about ten hours. Well done, team bacteria![28]

The Builders

As we are learning, bacteria do more than just talk, they also cooperate and like to build communities. There are more things that microbes are building or will be producing for us in the future with the help of synthetic biology, and we will dive deeper into just what this wonderful shadow workforce is up to in chapter 5. But first, it's important to understand how bacteria engineer their basic building tools of quorum sensing to create their own "cities"—which microbiologists call biofilms. To the human eye or touch or smell, biofilms are basically thin layers of slime that coat things like river rocks, the slippery lining of a pool, or the lining of our digestive systems, on implanted medical devices like a pacemaker, or the early morning covering on your teeth. Biofilms exist anywhere there is a source of water. Biofilms operate like any urban city: diverse species living side by side, creating food co-ops to share resources, consume each other's waste, and form a neighborhood watch to monitor external threats like antibiotics.[29]

Welcome to the Biofilm

Building a biofilm, escaping from the biofilm, or even living in a biofilm requires some sort of coordination among the millions of bacteria that make it up. They can do so by communicating with each other, using a chemical language of proteins. Jamming the bacterial language (although there are many) or interfering with their key factors required for coordination has proven to be a successful strategy to block or modify biofilm formation, at least in laboratory settings and some clinical pilot studies.

Likewise, repurposing the bacterial language has shown promise. For instance, when we co-opt the bacterial language to signal "escape from the biofilm!" we can trick biofilm bacteria into giving up their protective lifestyle and convert to planktonic cells again. The added benefit is the planktonic cells are more susceptible to antibiotics.

Controlling biofilms in the future will likely require a combination of strategies, addressing both attachment and escape, with and without the use of antibiotics and communication blockers, and likely in a manner more or less tailored toward the different bacterial lifestyles.

−Karin Sauer, professor of biological sciences, Binghamton University, State University of New York, via *The Conversation*[30]

The Survivors

One thing scientists have learned is that bacteria are good organizers, especially in times of stress. They are even known to swap genes between mixed species in their colony to resist antibiotics. But not all of their survival mechanisms are so tamely cooperative. Myxobacteria in particular are known to signal their fellow bacterium to soldier up and form predatory groups, known as swarms. Acting eerily like a wolf pack, myxobacteria move and feed cooperatively in their swarm. But mostly they like to dominate and take over a host. Usually this

host is a willing one, a cell displaying receptors to which bacteria adhere and then enter the cell. Other tricksters in the bacterial world include certain strains of *E. coli*, who are able to infiltrate a host cell even though that cell doesn't have the specific receptors that bacteria read. Instead, the nimble *E. coli* bring their own receptor that they can attach and use to enter the cell.[31]

While bacteria are often blamed for getting us sick, it's actually viruses that do the damage by infecting bacteria. Bacteria themselves get sick by being infected by a type of virus called bacteriophages ("bacteria eaters"). The viral contamination of bacteria occurs when the virus penetrates the bacteria cell's membrane, the outer shield that keeps the cell in place much the way skin wraps the bodies of humans and mammals. Some bacteria are able to change the shape of their cell walls to prevent viral attack, while others have what is called an "adaptive immunity," which is the ability to store memory of an attacking virus, so they can shift into defense mode when they recognize the perpetrator. Interestingly, humans have developed

Microbial Ninja Warriors: Bacterial Immune Evasion

A dark, shadowy figure makes its way through a well-fortified house, stealthily sidestepping traps, inactivating alarms, and taking down every guard that it encounters. This may sound like a typical scene from any ninja movie, but a similar scenario occurs whenever a bacterial pathogen successfully enters and infects our bodies. Much like a ninja, disease-causing bacteria, including *Mycobacterium tuberculosis*, *Neisseria meningitidis*, *E. coli*, and others, are all highly skilled in the age-old art of immune evasion.

Immune evasion strategies are those bacterial pathogens use to avoid or inactivate host defenses and ensure their own survival within a host. They employ tactics such as modulating their cell surfaces, releasing proteins to inhibit or degrade host immune factors, or even mimicking host molecules. Mastery of these camouflaging and precise weaponry techniques by bacterial pathogens significantly complicates efforts to develop new vaccines and innovative treatments.

—Erica Bizzell, excerpt from "Microbial Ninja Warriors: Bacterial Immune Evasion," American Society for Microbiology[32]

adaptive immunity as well (probably with the help of our bacterial roommates). Some viruses also adapt and change over time, tricking the memories of bacteria. A simple example of this is the common cold.

The Microbial Jump from Animals to Humans ... and a Little Thing Called COVID-19

Zoonotic diseases are cross-species in nature, traveling from animal hosts to infect humans. According to the World Health Organization (WHO), the world sees an estimated one billion cases of illness and millions of deaths every year from zoonotic diseases. All available evidence for COVID-19 suggests that it has a zoonotic source. And while we were under the pandemic of COVID-19, there were still other zoonotic-originated outbreaks happening: another Ebola spread in parts of Africa, Saudi Arabia saw a resurgence of cases of MERS (Middle East respiratory syndrome), and China quietly battled a few cases of a rare but deadly strain of hantavirus. The big question in the minds of epidemiologists is why, with all of our modern medical advancements, is there still this wildly contagious viral jump from animals to humans?

Thomas Abraham, an adjunct associate professor at the University of Hong Kong's Journalism and Media Studies Centre and author of *Twenty-First Century Plague: The Story of SARS* feels that "each time you have a disease, it is a sign that deep down the relationship between man and microbe has changed in some fundamental way." In his book, he explains that with the exponential growth of the human population, we are having to expand into areas humans haven't lived in before, claiming more forested terrains, or draining lakes for new buildings, exposing humans to new forms of animal life, and most likely the pathogens they carry.

In an interview with the *Guardian*, the Danish environmentalist Inger Andersen, Executive Director of the UN Environment Programme, said that "nature was sending humankind a 'message' through the coronavirus pandemic and climate crisis," stressing our natural system, and perhaps our own immune systems.

—LIVEMINT[33]

Give CRISPR Thanks and Gratitude to Bacteria!

If you've heard of CRISPR—the new biotechnology that allows us to cut and edit genes—then you owe a debt of gratitude to bacteria's "memory" immune response. CRISPR stands for "clustered regularly interspaced short palindromic repeats," and is a family of DNA sequences found in the genomes of bacteria (and archaea, but for now we're going to focus on bacteria). Essentially, scientists discovered a pattern of short, repeated sequences in bacterial DNA that had some gaps in between the repeated cycles. Those sequences are left over from DNA fragments of previous bacteriophages that had infected the bacteria and are the sequences bacteria recall for immune response. What they do is pick up the leftover fragments from their invaders and use it to detect and destroy DNA from similar bacteriophages during subsequent infections.[34] It's like a superpower where the bacteria being invaded by a virus goes back into its DNA "data bank" to use the fragments from a previous virus to cut up

the new virus before it ever has a chance to copy itself and multiply. CRISPR is bacteria's super "cut and paste" power—cut the invading virus, copy parts of it, back it up, and then paste it back when needed to fight off the next virus.

The adaptive immunity powers of the bacteria involve RNA-guided Cas proteins to enable bacteria genes to edit and cut foreign pathogenic DNA and RNA. For the geeks like us who want to know, the simple way to understand the functional difference between DNA and RNA is that where DNA replicates and stores genetic information (it's the blueprint for all genetic information within an organism), RNA is the converter, transforming the genetic information contained within DNA into a format that can be used to build proteins and more. Together, they are the most important molecules in cell biology.[35] They're also the key to bacteria's adaptive immunity, working in tandem to not only cut off pathogens but relay the intel on these pathogens to the daughter cells after the "mother"

cell divides.[36] Whereas humans have a complex immune system that involves several specialized cell types, bacteria rely on this storage system and the passing of information from DNA via RNA to stop the virus in its tracks.

While bacteria have practiced gene editing for eons, we have only just begun to tap the possibilities of CRISPR, and, of course, to consider the ethical questions about how we use this biological tool. There is a lot of optimism for its potential to support new health and sustainability solutions, including the engineering of biotechnology products with greater resistance and efficacy, and in the treatment and prevention of diseases.[37] And we have bacteria to thank for showing us the way!

Endospores Are One of Nature's Most Resilient Means of Surviving

The Shapeshifters

Some bacteria are known to transform themselves into endospores (a dormant, tough, and non-reproductive structure) to resist heat and dehydration, especially during times when nutrients are scarce. It's an ingenious way that certain bacteria avoid suffering or even death by turning themselves down to just the essentials and hibernating. To induce this shutdown mode, the bacterium divides within its cell wall, and then basically one side totally engulfs the other side.[38] What is even more fascinating is that this endospore state enables bacteria to hole up for years, even centuries. There are revived spores that have been thought to be dormant over ten thousand years, and scientists have been able to revive some spores after millions of years of dormancy. One report claims to have found in salt crystals viable spores of *Bacillus marismortui* that are 250 million years old.[39]

To be clear, not all bacteria have the superpower to shapeshift into endospores. In case you were wondering who the lucky ones were, they include *Bacillus cereus* (commonly found in soil and food), *Bacillus anthracis* (the agent of anthrax!), *Bacillus thuringiensis* (used as a biological pesticide), and *Clostridium botulinum* (produces the neurotoxin botulinum, the most potent toxin known to mankind), and *Clostridium tetani* (that causes tetanus).[40]

The Movers and Shakers

Navigation, exploration, and the hunt for food are not unique to humans or animals—some species of bacteria possess these drives as well, which we are still learning about. The *E. coli* bacteria in our digestive system, for example, spend some of their time traveling around and visiting different environments there, encountering the sugar lactose as it tries to find the related sugar maltose (*E. coli* love to feed on sugar).[41] It's a behavior known as chemotaxis, where bacteria move when there is the presence of

chemicals in their environment. *E. coli* that live outside the human body (and basically on almost every surface) get moving when just a few molecules of food are dropped in their colony.

The bacteria known as magnetotactic bacteria have special organelles with magnetic crystals that help them align to the Earth's magnetic field, much the way birds, whales, and other migratory species do.[42] There is a species in magnetotactic bacteria called *Magnetococcus marinus*, or the "microswimmers" who use two bundles of threadlike flagella in addition to their intracellular "compass" to guide intentional movement or a specific route. What makes *Magnetococcus marinus* especially attractive to science is the notion that you can potentially control their movements using a magnetic field—which is precisely what an international team of researchers from the University of Göttingen, the Max Planck Institute of Colloids and Interfaces in Potsdam, and the CEA Cadarache (France), discovered when they determined their average swimming speed. The importance of this discovery is that *Magnetococcus*

The Synchronized Swimmers

To explore their environment, individual bacterial cells swim in stretches for a duration of seconds, interrupted by short reorientations (tumbles). While swimming, the bacteria use a minimal sense of smell to measure changes in their chemical environment, such as nutrients, to steer their swimming partners toward better life conditions. The network of proteins that controls this behavior uses a chemically encoded memory that helps bacteria to compare the current conditions to the one experienced a few seconds prior.
—*Nature Communications* [43]

marinus can be used as a biological model for medical microrobots that move around in our blood to find and target tumors, for example. Or, do we dare say, genetically engineering them one day to *be* the microrobot? [44]

And one more example of micromotility, and this one is a colossal one in terms of microbial size: the icky yet captivating behavior of the slime mold known as *Physarum polycephalum*, a colony of amoeba-like organisms that always inches its way to the shortest route through a maze. Despite not

ure out a maze when there is food at one end. While this hungry mold is not as scary or massively threatening as the lead character in the 1958 classic *The Blob*, researchers are still figuring out its keen sensory system.[45] (Feeling another remake here, anyone?)

The Altruists

We've learned so far that microbes, and in particular bacteria, are smarter than we think, have a memorable adaptive immunity, can work their way through food—and also that they do this as a group, team, or social community. Being super-social microorganisms, bacteria possess a type of altruism not dissimilar from our own need to consider the benefit of others over ourselves. Some may describe this as the foundation of our social contracts.

One way to see the altruism of bacteria is in how they build their biofilm "cities," emerging through the collective behavior of thousands or millions of cells.[46] When under stress for nutrients,

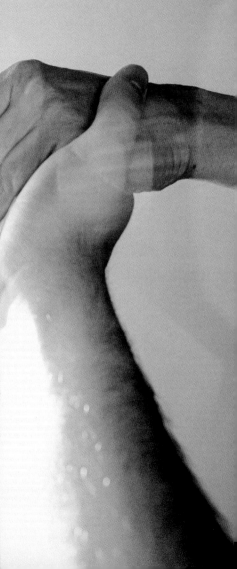

some bacterial colonies organize themselves in a way that helps to maximize nutrient availability. *Bacillus subtilis* is one species that uses electric signals (ion transmission) to synchronize growth so that the innermost cells of the biofilm do not starve.[47] Other animals practice a similar act of selflessness. For example, when a group of meerkats are out foraging for food, one selfless individual stands watch for predators, giving up its own feeding time for the greater good of its group—an example of what biologists call altruism.

In his 1976 book *The Selfish Gene*, the evolutionary biologist Richard Dawkins theorized about a special type of gene that may be the reason why organisms direct selfless behavior toward other organisms, which he identified as the "greenbeard gene." There has been a recent discovery of the so-called greenbeard genes in microbes, and the social behavior of bacteria, fungi, algae, and other single-celled organisms, which could help us understand the origins of altruism.

Get Social!

Microbes and Their Greenbeards

Nature has a magical way of ensuring certain traits and characteristics are replicated in the next generation of an organism, and this is done through genes. From humans to the single-cell organism, we each inherit genes from our parents, or in the case of prokaryotes, which include bacteria and single-celled archaea, their chromosomal DNA is passed on to offspring asexually.[48] In a human family unit, siblings share about 50 percent of the same genes, half from each parent. Some of these genes are the ones that promote altruistic behavior. There's a biological theory of altruistic gene expression based on siblings: if a gene for altruism encourages a sister to help her sister, then the gene "thinks" there is a 50 percent chance that the sibling has a copy of the same gene. It doesn't need to recognize itself to express its trait; it simply needs to help its kin.[49] This biological explanation for altruistic behavior is what some biologists call "kind discrimination," and Dawkins's greenbeard theory questions if

there might be another way that altruism is expressed in nature, one that wouldn't involve the direct exchange between relatives. According to his greenbeard system, organisms must acquire three things: an arbitrary peculiarity that acts like a signal (the greenbeard), the ability to detect the greenbeard on others, and the tendency to treat such greenbearded individuals preferentially.

One form of altruism is "kind discrimination," which in the study of bacteria microbiologists define as a "mechanism in which individuals preferentially help or punish others." In many situations where altruism occurs, there is a need that is expressed by one party, or a group, for which help is required. Microbes have situations where they need each other's help, and elicit a response, particularly when there is a lack of food. The social amoeba *Dictyostelium discoideum*, a soil-dwelling species commonly referred to as a "slime mold," crowd together with thousands of other amoebas and, when necessary, some of these single-celled organisms sacrifice themselves to form a stalk or ladder that stretches skyward to help others climb up and disperse to find a new food source. This starvation state causes the amoebas to stick together, a cell-to-cell adhesion that helps them amass into a stronger form that searches for food, while sticking copies of itself on the other cells traveling in the hunt. This procession is a form of kind discrimination, as it helps those they recognize, yet excludes the cells that don't match the others in the group.[50]

Another example of kind discrimination among social microbes has been found in a fungus that uses the greenbeard system to decide whether or not it should approach a new individual in the neighborhood and fuse with it. This can also happen over a distance. Researchers from the University of California, Berkeley, discovered that the filamentous fungus *Neurospora crassa* uses a set of highly divergent genes to discriminate "self" from "non-self" cells to actively seek out those favored cells (those of the same "kind") that were located a distance beyond their immediate environment. How do they do this? It might be due to the way they communicate. When genetically identical asexual spores of *N. crassa* germinate (termed germlings), their growth is guided by an outside chemical stimulus (aka chemotropic interactions), and what ensues is eventual cell fusion. A dialogue occurs between the germlings: the genetically identical cells alternately "listen" and "speak" (like a microbial game of Marco Polo). The researchers found that the *N. crassa* populations had discrete communication groups, where their germlings from the same communication group were chemically attracted to each other, while the germlings from different communication groups grew past each other to find a germling that matched their com-

Kindness in Nature

munication type. But what is most interesting about this discovery is that there are highly variable genes called "doc" genes (for "determinant of communication") within the fungus populations that can sift through the various communication groups in a population to identify those that have a similar "self" identity. The doc genes function as greenbeard genes, involved in mediating long-distance kind recognition by actively searching for one's own type, which results in cooperation between non-genealogical relatives.[51]

Science is just beginning to study the phenomenon of greenbeard genes and we are still not sure how widespread this altruistic act of connection and kindness is in nature. We are also still learning how "family" is essential to keeping the greenbeard genes going—in microbes and potentially other species, the family group is essential for survival. There are even some hypotheses that the way bacteria defend themselves and their colony from viruses is an example of altruistic behavior as well.

So, here's a question that we may not have the answer to yet, but gives us pause: since we humans are superorganisms made up of trillions of microbes, could it be that we are socially altruistic because of bacteria? Just asking . . .

Symbiotic Relationships: Meet the Holobionts

It's Not Us vs. Them.
It's Us + Them.

You—yes, *you*—are unwittingly an altruistic microbial host. As it happens, just about every living organism is a host, and we carry around with us a multitude of microbial associates, communities, and megacities. These microbial colonies are what a body of research has defined as the "microbiome"—the microorganisms in a particular environment (including the human body as a whole or a part of the body). We are still decoding their genetics and discovering more about the menagerie of microbes that dwell on and within us, but every day scientists learn more about these remarkable organisms. What we do know for sure? You are never alone. Ever. Microbes exist everywhere.

The combination of a host and all of the resident microbes that live in it and on it is called a holobiont. As we've learned, these communities can be mutualistic (where both organisms benefit) or commensal (where the relationship is neutral, with no meaningful benefit to either). And biologists have learned that even while living in a holobiont, some of the microorganisms still express their behavior of excluding or killing others (with toxins) or releasing enzymes and nutrients to the benefit of their neighbors. Their survival and altruistic instincts make them even more beneficial to us humans and the other hosts they tag themselves to. It's not microbes vs. us. It's us + microbes for the ultimate game of life.

The Hologenome Is Greater Than the Sum of Its Parts

In 1991, Lynn Margulis, the legendary evolutionary theorist and biologist who popularized the theory of symbiosis and Gaia, first coined the term "holobiont," drawing from interest in long-lasting and tightly integrated associations made between two "species." She thought that the coming together of two species to form a new symbiotic consortium was analogous to an egg

and a sperm of two organism "partners" coming together to form a new organism. But Margulis challenged that this life cycle doesn't just involve two species; rather it is a host of organisms that live together as one, and instead of sending out tiny cells to reproduce, holobionts send out individual organisms of different species—those that live harmoniously with the host and all its microorganisms.

Coral reefs are convenient examples of holobionts. The most beautiful coral reefs in the world—from Australia's Great Barrier Reef that spans over 1,400 miles (2,300 kilometers) to the Belize Barrier Reef, to the Red Sea Coral Reef—are holobiont environments, and their health is due to several microbe colonists and cohabitators. Coral reefs are built by thousands of tiny animals called coral "polyps" that reproduce by cloning themselves, working side by side with photosynthetic algae that they host within their own cells. Both the coral polyps and the algae are harmoniously dependent on each other: the algae provide nutrients to the coral; the coral gives the algae protection. But the coral needs more than nutrients, it also needs immune protection from disease, which the algae can't help with. So coral selectively extends an invitation to helpful bacteria to help crowd out the harmful ones, expanding the "guests" that live within the reef. Essentially, the coral reef is a host in which several microbes are annexed so everyone can thrive.

The coral reef as a host is a simple way to imagine the evolution of Margulis's seminal theory of holobionts into a new hypothesis that can be likened to a "conglomerate" genome, known as the *hologenome theory of evolution*. The hologenome theory suggests that the host and all its resident microbes (the holobiont) is made up of the host genome plus all the microbial genomes, forming a hologenome. A key implication of this theory is that natural selection acts on the whole hologenome of the holobiont, and not just on the genome of its individual organisms. The hologenome is therefore greater than the sum of its parts.[52]

Are You and Your Microbes a Community or a Single Entity?

Derek Skillings, biologist and philosopher of science, explored the question of community vs. individual in his September 2018 article in *Aeon* based on the evolved theory of the holobiont. He challenges the traditional view in medicine that treats the human body (and perhaps that of animals) as a battleground of good vs. bad microbes. If you adopt a contemporary, ecosystem-based perspective, it's clear that we need to consider humans as hosts of a microbial habitat to be managed and balanced, perhaps requiring the enticement of competition between different species of microbes to let the good prevail. In this sense, what "counts as 'healthy' can depend on what kinds of services we want out of our attendant ecosystem," explains Skillings, suggesting that we "might expect stable coadapted partners living in concert across holobiont generations."

Like any scientific theory, thinking about the holobiont will undoubtedly evolve in the coming years, especially as we decipher more of the genome and microbiome and the environmental effects that shape the exposome of, well, everything. But for now, Skillings suggest we adopt an expanded version of the "us versus them" picture of medicine. After all, there are more of "them," and many are our allies. We need to take care of them, just as they help care for us.[53]

Why, the very cells that build you are themselves colonies of bacteria, replaying the same old tricks we bacteria discovered a billion years ago. We were here before you arrived, and we shall be here after you are gone.

—Richard Dawkins, from *The Ancestor's Tale: A Pilgrimage to the Dawn of Evolution*[54]

We Are Family

I've Got All My Microbes with Me!

Trillions of microbes affect every stage of our life—from birth to old age. Now that you know your body contains many more microbes (including bacteria and viruses) than human cells, it's time to play nice with your new friends.

Welcome to Your Microbiome!

Your human development and all the systems in your body have evolved, or coevolved, with your microbes. Small but super-packed viral and bacterial genomes are teeming with information about your health. Your gut and brain are connected physically through millions of nerves, and the gut and its microbes also control inflammation and make many different compounds that affect brain health. This brain/immune/gut connection is what researchers call "The BIG Axis." Various parts of the body have very different communities of microbes, yet the distinct microbes in our gut, our oral cavity, and on our skin *all* turn out to be critical to our microbial fingerprint that can identify the unique you. According to a study by Harvard's T. H. Chan School of Public Health our personal microorganisms have distinguishing features that stay with us over time, and our microbiome data may be able to identify us like DNA does in genetic tests.[1]

Microbiomes differ from person to person (and body part to body part!), a discovery that has birthed a new industry of microbiome companies that strive to offer precise food, supplement, and even drug recommendations based on an individual's unique gut microbiome.

Every day it seems new evidence arises tying the microbiome to an array of conditions, from anxiety and obesity to sleep disorders. Even Big Pharma is exploring how the microbiome is linked to how patients respond to drug treatments such as chemotherapy. Many research hours have been spent trying to crack the microbiome's complexity, and in this chapter we'll take a look at the promising yet challenging work being done to decode the microbiome and transform personalized preventive care today, and explore the frontier to see what we will do in the near-to-far future to keep our microbiome in good shape.

From "Survival of the Fittest" to "Fitting in" with the Rest of Life

In 2014, the Secret Science Club at the Bell House in Brooklyn, New York, hosted Dr. Martin J. Blaser, the then-director of the Human Microbiome Project at New York University (he is now the Henry Rutgers Chair of the Human Microbiome at Rutgers University) and author of the seminal book *Missing Microbes: How the Overuse of Antibiotics Is Fueling Our Modern Plagues* for an evening discussion titled "It's a Jungle in There" about the human microbiome. We were lucky to be in attendance as Dr. Blaser shared his thirty years of knowledge on the role of bacteria.

Dr. Blaser's talk opened our eyes to the fact that parts of our microbiome are starting to disappear. Ordinarily, each of us consists of an estimated 30 trillion cells but are host to more than 100 trillion bacteria and fungal cells. Some microbes have specialized properties and live and thrive in various areas of our bodies such as our skin, mouth, nose, gut, ears, and, in women, vagina. We depend on them to digest our foods, regulate our moods, support our immune response, and more; their health is our health.

But thanks to our cultural obsession with everything from antibacterial soap and hand sanitizers to the overuse of antibiotics to treat disease (both our own and those of the animals we consume), the quantity and variety of microbes we carry is facing a kind of ecological Armageddon.

Antibiotic overuse causes devastation to our digestive system by erasing the diversity and complexity of microbes in our gut, destroying the helpful, friendly bacteria along with the bad ones. A similar scenario occurs when we try to disinfect, or worse sterilize, our skin and oral cavity. Treatments such as the severe scorched-earth antibiotic approach that kills everything in sight are being rethought. We are moving from the "survival of the fittest" to "fitting in" with the rest of life. It's not a winner-takes-all approach of us vs. microbes; it's a mutuality where we both need to coexist and work together toward being beneficial for each other.

Why You Should Be Worried about Missing Microbes

On his website, Dr. Martin Blaser shares critical insights on his hypothesis of how the overuse of antibiotics, C-sections, and antiseptics has permanently changed our microbiome, causing an increase in modern diseases such as obesity, juvenile diabetes, and asthma.

—missingmicrobes.com[2]

Healthy Gut Microbiome Reference Database

GutFeeling Knowledge Base's goal is to map the healthy gut microbiome so that they and other researchers can use the data to develop disease-specific prediction models. Having this overarching umbrella in place would be the next step toward better understanding how humans interact with their microorganisms.

—George Washington School of Medicine & Health Sciences[3]

It should be clear by now that even short-term antibiotic treatments can lead to long-term shifts in the microbes colonizing our bodies. A full recovery or bounce-back of healthy bacteria is in no way guaranteed, despite the long-held belief that such was the case. But that is not my only worry. I also fear that some of our residential organisms—what I think of as contingency species—may disappear altogether.

—Martin J. Blaser, *Missing Microbes: How the Overuse of Antibiotics Is Fueling Our Modern Plagues*[4]

Data, Data—And More Data

The microbiome could someday help physicians treat patients for a medical condition before symptom onset. Just like a diabetic who checks their sugar levels, we would hope that, in the future, people could check their microbiome and see if they are getting into a state of microbiome shift.

—Julia Segre, PhD, a scientist looking at atopic dermatitis flares in children as part of the NIH federally funded Human Microbiome Project (HMP) project via Healio[5]

Probiotics to the Rescue

Poised to reach over US $38.7 billion by the year 2025, probiotics will bring in healthy gains adding significant momentum to global growth.

Probiotics harness the huge potential of the human microbiome in transforming health and wellness and are the preferred ingredient for digestive health-related food and drinks.

Probiotic bacteria are growing in prominence in functional and optimal health foods.

Probiotics are emerging as leading mood-enhancing and brain health aids.

—ResearchAndMarkets.com[6]

Shifting from "Anti" to "Pro"

Probiotics Will Be to the Twenty-First Century What Antibiotics Were to the Twentieth Century

Your gut is home to the microbiome, an army of microbes that influences your mood, weight, and immune system. Junk food and overuse of antibiotics have wiped out many "good" gut bacteria, leading to a modern plague of allergies, food intolerances, and obesity. By nurturing our microbiomes, we may be able to reverse some of these devastating changes.

The Living World of "Biotics"

Before we continue exploring how to keep our microbiome in good shape, we need to get acquainted with "biotics." We've all heard about probiotics (yogurt anyone?). We know they're good for digestion, but what exactly are they? And what role do all the other "biotics" coming into the wellness market play? How do they help us?

Probiotics

Probiotics, "good bacteria," are live microorganisms that keep your body healthy. Probiotics live in our stomachs and their primary function is to help us digest food, though probiotics like *Lactobacillus* (which is also the most common probiotic) help strengthen the skin microbiome and provide relief for various sensitive skin conditions. Foods with probiotics include yogurt, kefir, sauerkraut, tempeh, kimchi, miso, kombucha, pickles, traditional buttermilk, and natto.

Prebiotics

Prebiotics feed the good bacteria (probiotics) in your gut. By consuming fiber, which is the basis for prebiotics, you can fuel and reenergize your gut's bacteria. Topically, prebiotics recalibrate the microbiome by fortifying the skin's beneficial microflora, promoting healthier skin. Foods with prebiotics include chicory root, dandelion greens, Jerusalem artichokes, garlic, onions, leeks, asparagus, bananas, barley, oats, apples, konjac root, cocoa, burdock root, and flax seeds.

Synbiotics

Synbiotics are the combination of prebiotics and probiotics, usually in the form of supplements that help probiotics pass through the intestinal tract. However, there isn't much data to support if a supplement works better than just supporting taking probiotics and eating a diet rich in prebiotics.[7]

Postbiotics

Postbiotics are by-products of the fermentation process carried out by probiotics. In simpler terms, probiotics feed on prebiotics and postbiotics are produced. Postbiotics are responsible for multiple important health-boosting functions in our gut, supporting the immune system to help fight infection. Some examples of postbiotics include organic acids, bacteriocins, carbonic substances, and enzymes. They result naturally from the existence and survival of microorganisms living in our gut, though they can also be added directly through therapeutic processes. Topically, postbiotics help regulate the composition of your skin's natural bacterial ecosystem.[8]

Psychobiotics

Psychobiotics are pre- and probiotics that affect the gut-brain axis and can aid in mood regulation. Psychobiotics are live organisms that, when ingested in adequate amounts, are capable of producing and delivering neuroactive substances such as gamma-aminobutyric acid and serotonin.[9] Relatively new to the biotics family, psychobiotics are still being researched for the possibility of using them to treat depression and regulate anxiety.[10]

Learn from
Your Gut!

Microbiome tests attempt to detect the presence of different species of microorganisms in a fecal sample. The test results give medical professionals key clues about the diversity of the gut microbiome and how it compares to that of others. There's preliminary evidence that analyzing the gut microbiome in a stool sample can help predict who will do well on a certain diet.[11] One day, we may understand how combining information about our microbiome with other diagnostics such as genomic tests might provide insights into how a person's cellular health, immune system health, gut health, stress response, and biological age may change over time. This data may translate into actionable health markers, which can be improved by following personalized nutrition recommendations.

Biotechnology companies such as Day Two, Kallyope, TargEDys®, and Viome are working to unlock the therapeutic and nutritional potential of the gut-brain axis and develop precision medicine therapies for a variety of metabolic diseases. Other companies, such as Pendulum and PanTheryx are focusing more specifically, developing therapies to metabolize glucose to manage type 2 diabetes and employ the microbiome in colostrum to treat a variety of gastrointestinal conditions, respectively.

Most companies ask you to send in a stool sample, which you take yourself and post in a secure package to a laboratory to analyze the results. Today, curious minds (or should we say curious guts?), are contributing to microbiome data that could become central to personalized health.

We finally have the technology available to be able to digitize the human body at a molecular level and analyze system-level biochemical activities. This deep understanding allows us to recommend to each individual why they should eat certain foods, and why they should avoid certain foods based on their own individual biology, with the goal to prevent and reverse chronic disease. This is bringing us one step closer to our mission to make chronic diseases a matter of choice and not a matter of bad luck.

—Naveen Jain, CEO of Viome[12]

Listen in on the Cross Talk on the Brain, Immune System, and Gut Axis

All disease begins in the gut.

–Hippocrates of Kos (c. 460–c. 370 BCE)[13]

The BIG Axis

In chapter 1 we established the human being is a superorganism. Before we dig deeper into the microbiome it is important to understand that our superorganism's biological processes actually talk to one another. This cross talk is happening all day between the brain, immune system, and gut. This connection is called the BIG Axis. And, like the word implies, it's a pretty BIG deal! The effects of gut microbiota on the brain and behavior are fulfilled by the microbiota–gut–brain axis, which is mainly composed of the nervous pathway, endocrine pathway, and immune pathway. In simpler terms: if the gut is healthy and the immune system is working, our brain is happy too. It is a simple conversation.[14]

However, when the gut microbiota gets muddled the cross talk between our body and mind can become uneasy or dis-eased. So how does this two-way communication work? It is a biochemical signaling that flows between the gastrointestinal tract (GI tract) and the central nervous system (CNS), and the gut's microbiota produce a range of neuroactive molecules essential for the process. These neuroactive molecules include acetylcholine (the chemical that motor neurons of the nervous system release in order to activate muscles), melatonin (a hormone that regulates the sleep-wake cycle), and serotonin (a neurotransmitter that modulates mood, memory, and learning)—all essential for regulating sensation in the gut. Changes in the environment of our gut flora due to diet, drugs, stress, or disease also affect changes in levels of circulating cytokines (peptides important to cell signaling), some of which can affect brain function.[15]

The Second Brain in Your Gut

That "gut feeling" is for real—and your gut microbes may play a role in it, helping with the signaling of your "second brain," which we call the enteric nervous system (ENS) of your belly. While your gut brain can't solve the *New York Times* Sunday crossword, it does directly affect your mood, your decisions, and stress level. That's because the enteric nervous system has two layers of more than 100 million nerve cells that line the entire gastrointestinal tract. Running from the esophagus down to the rectum, this superhighway of nerve cells forms a feedback loop of chemicals and hormones.[16] Your gut microbial environment influences the release of the neurotransmitter serotonin, which is important for regulating your feelings of happiness, and dopamine and GABA, other neurochemicals associated with moods. They are transported through the vagus nerve, part of the superhighway, and can influence the brain to tune its behavior to the messages it receives from the community of bacteria in the gut.[17]

On the flip side, stressful situations can induce changes to the gut microbiome, causing the release of defensive molecules called inflammatory cytokines, which cause gut inflammation, disrupt brain neurochemistry,

and make you more vulnerable to anxiety and depression.

Scientists are still untangling the psychological and emotional effects gut bacteria have on our well-being, but early research has shown that tweaking the balance between beneficial and disease-causing bacteria in the gut of mice can alter the animal's brain chemistry. The changes lead the mice to become either bolder or more anxious, and when faced with even mild stress, the microbial balance in the gut was disrupted.[18] These findings offer the possibility of using beneficial (or probiotic) bacteria to treat mood and anxiety disorders—either through the treatment of beneficial microbes themselves in the body, or by drugs that mimic the metabolic functions of good bacteria. Just imagine one day being able to take a mood-enhancing microbial cocktail in lieu of Prozac (or your daily glass of wine).

While we are still in the infancy of microbiome research, understanding the BIG Axis will undoubtedly help us heal. Various microbiota-improving methods including fecal microbiota transplantation, probiotics, prebiotics, a healthy diet, and healthy lifestyle have shown the capability to promote the function of the brain, immune system, and gut axis. In the future, it will be possible to harness the gut microbiota to improve brain and mental health and prevent and treat related diseases.

The Gut's Pavlovian Response

You have probably heard of the expression "Pavlov's dog" in describing a conditioned response to a trigger. Ivan Pavlov's body of research in the early twentieth century demonstrated the first known brain-gut interaction during digestion, when the release of gastric and pancreatic secretions responded to sensory signals, such as the smell and sight of food.[19]

Analyzing Gut Microbiota in Pregnancy

Because pregnancy is a time marked by changes in a woman's body, such as immunological changes that can influence intestinal flora, the Bioscience Institute conducts an intestinal and vaginal microbiome analysis through their MICROBalance platform. According to the Bioscience Institute, some gut microbiota changes are good for the mother and child's health, whereas others could be associated with pregnancy complications or compromise the child's gut microbiota development. After the first trimester, levels of *Faecalibacterium* (a bacterium with anti-inflammatory activities) and the diversity of microorganisms decrease, whereas some health-friendly bacteria may increase and play an important role in defending against pathogens, reinforcing the intestinal barrier, and nutrient metabolism.

—Bioscience Institute[20]

Mom, Microbes, and Me

As we emerge through the birth canal, eyes and mouth wide open, we get our first dose of microbial flora from mom. During pregnancy, a mother's microbiome adjusts to an optimal microbial mix for her baby. This bacterial elixir starts the whole process. Our gut microbiome changes quickly over the first year or two, shaped by microbes in breast milk, the environment, and other factors, and stabilizes by the time we're about three years old.[21]

If you are born by C-section, your microbiome development starts differently. Studies published by the NIH suggest that these differences could be one of the reasons why babies born by caesarean have a higher risk of conditions including asthma and type 1 diabetes. There has been a controversial birthing trend where newborns delivered by C-section are given a smear of microbes from the mother's vagina soon after birth. Called "vaginal seeding" or "microbirthing," the idea is to help make C-section babies more immune to disease

Intriguing as this trend seems, the American College of Obstetricians and Gynecologists (ACOG) opposes this practice, citing concerns that the baby could develop life-threatening infections during the seeding.[22, 23]

Skin-to-Skin Contact

Kangaroo Care

According to Mercy Health, skin-to-skin contact (also known as kangaroo care) is when a naked baby is placed directly on the bare chest of a mother or caregiver as soon as possible after birth. Hospitals and birthing centers are finding ways to keep mom and baby together, whether the baby was delivered vaginally or via C-section, so they can enjoy the many benefits of kangaroo care.

Skin-to-skin contact is important in baby's first moments, especially right after birth when about 10 percent of a baby's good bacteria comes from the skin around the mother's breasts—a big microbial boost for C-section babies as well. Breastfeeding plays a big role in populating an infant's

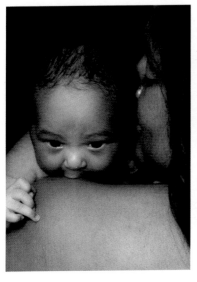
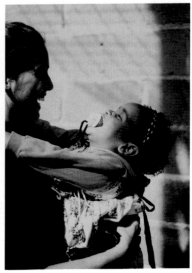

gut microbiome, with breast milk planting baby's tummy with 30 percent beneficial bacteria. It's also recommended to put off giving your baby its first bath until twelve hours after it is born to help nurture their skin's microbiome.[24]

Kangaroo care is not just between newborn and mom—it's a way for dad and the whole family to bond with the baby, while sharing some good family microbes. Some of the benefits include calming the baby, helping stimulate breastfeeding, stabilizing heart rate and body temperature, and improving sleep.[25]

Saliva Is a Window into the Body

The Human Oral Microbiome

There's a garden of over seven hundred species in your oral cavity, which makes your mouth the second largest residence of microbiota after the gut. Your mouth nurtures microorganisms that include bacteria, fungi, viruses, and protozoa.[26]

In 2010, microbiologist Floyd Dewhirst of the Forsyth Institute in Cambridge, Massachusetts, published in the *Journal of Bacteriology* a comprehensive examination of mouth-dwelling microbes,[27] which formed the basis for the Human Oral Microbiome Database—a resource that's now updated nearly every day.

Researchers are out to prove that the mouth may be the gateway to our health. In fact, a mom's saliva can strengthen her baby's immune system: studies show that sucking their infant's pacifier is associated with a reduced risk of allergy development and an altered oral flora in their child.[28]

The mouth may one day be considered equal to the rest of the body in its usefulness in diagnosing health. A phone-sized point-of-care saliva test can measure proteins that may indicate a developing disease in the mouth or elsewhere in the body. In the future, we may value microbes over medicine.

Salivary diagnostics is something that will probably be used much more in the future. Most people would rather spit into a tube than have blood drawn. And most everything that's in the blood is also in saliva. One of the things we're finding is that things you wouldn't think you could diagnose, not from the proteins and stuff in saliva but by the bacteria that live in our saliva; [for example,] pancreatic cancer seems to have a different profile of bugs in the mouth than someone without it. So even systemic diseases are reflected in the bacteria that are present in the mouth. In the future, a lot of systemic diseases will be able to be diagnosed using oral samples, saliva samples.

—Floyd Dewhirst, DDS, PhD, professor, Department of Oral Medicine, Infection and Immunity, Harvard School of Dental Medicine, Sputnik Futures interview, 2014

Are Microbes Affecting Our Weight?

Ever notice a friend who just never gains weight, or gains it slower than you? We used to chalk it up to genetics or a good metabolism, but now researchers are questioning whether our gut microbiome development plays a role. Science has already evidenced that an altered gut microbiome can lead to changes in metabolic function and, as a result, cause obesity. When it comes to being overweight, there are several ways gut microbes might influence the situation: through appetite, production of gases, efficiency of food usage, impact on the immune system, and inflammation.[29]

Researchers at the Children's Hospital of Philadelphia (CHOP) are observing the gut microbiome's development, starting with the first hours of infancy, in a two-year study. The researchers hope is to determine how the development of the gut microbiome may influence excess weight gain. A baby's gut will hold hundreds of

different species of bacteria, but at birth, there might only be ten or fewer species. Understanding why those particular bacteria are the first to populate the gut and what they're doing in those first hours of life may help develop a critical baseline of the gut microbiome.

The researchers focused on three species of bacteria in particular whose populations are known to be greatest in babies—*Escherichia coli*, *Enterococcus faecalis*, and *Bacteroides vulgatus*—and analyzed the genomes of these bacteria to determine why they are growing in infants. The bacteria didn't emerge in detectable concentrations until the infants were about sixteen hours old, and analysis showed that multiple strains of each bacterium were already emerging. Eighty-eight infants involved in this two-year study are African American, because of the growing childhood obesity concern in this cohort.

The researchers will continue to follow these healthy term infants to see how long early strains of bacteria linger, and, more impor-

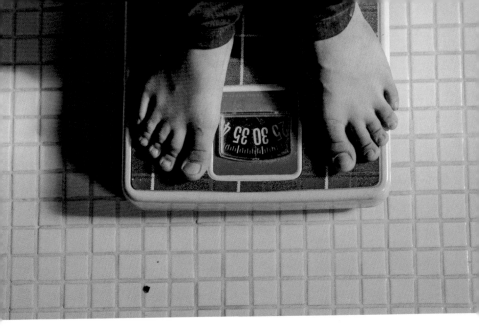

tant, learn what a normal growth pattern of gut microbiota looks like. The goal in the future is to enable a microbial intervention to prevent adverse effects on children, such as the development of diseases, when their microbiome changes.[30]

Obesity is a global epidemic, with more than one-third of the world's population either overweight or obese, a group that has tripled in size between 1975 and 2016, according to the World Health Organization. In 2019, 38 million children under the age of five were overweight or considered obese.[31] Obesity has been associated with food choices and lack of exercise, which are usually assumed to be conscious choices made by the individual. But recent studies have challenged the notion that obesity is a choice or result of our own actions.

Researchers have found that there is a difference between the gut microbiome of obese and non-obese individuals, indicating that obesity isn't just a matter of food choices or how the body uses energy from the food consumed. It is instead a matter of how gut bacteria process that food. This shifts how we look at obesity, as a disease rather than a consequence of poor food choices.[32]

We've learned so far that the gut microbiome affects different aspects of our body like our metabolism, mood, and immune system. As we develop, several key factors influence the composition of the gut microbiome in each individual. Besides the microbes we inherit from our mother's body, our diets and our lifestyle directly impact the composition of our gut microbiomes. This partially explains why siblings born from the same mother (who presumably start out with similar microbiota) can have different gut microbiomes.[33]

Considering that the gut microbiome plays a key role in digesting the food we eat and absorbing and synthesizing nutrients from our food, researchers are looking further into the link to obesity. One study exploring this link examined the gut bacteria in obese and lean mice and human volunteers. Interesting fact: although the majority of mouse gut species are unique, the mouse and human microbiota(s) are similar at the division level of microbes. The research team, led by

the Center for Genome Sciences & Systems Biology at Washington University School of Medicine, St. Louis, Missouri, found that the relative abundance of two dominant types of bacteria, *Bacteroidetes* (rod-shaped bacteria that are widely distributed in the environment, including in soil, seawater, and also in the guts and on the skin of animals) and *Firmicutes* (who play an important role in beer, wine, and cider spoilage), was different in obese and lean subjects. Investigating further, the researchers found that changes in the gut microbiome have a significant impact on metabolic function. The microbiome in obese mice could harvest more energy from food than that of their lean counterparts, impacting the amount of body fat. They also found that colonizing lean mice with microbes from obese mice significantly increased total body fat. The result shows that the presence of a gut microbial community may affect the amount of energy that is extracted from the diet, leading to the obesity of the host.[34]

There are several other studies that also show the gut microbiota's response to weight loss and

the effects of the gut microbiome on insulin resistance, inflammation, and fat deposition in the body.[35]

One of the more intriguing insights researchers are expanding on is how the extensive diversity of these microorganisms in our gut results not just from our diet and lifestyle but also from our geographical location, infections, sex, age, and genetic background, and their effects on everything from metabolism to immune response to behavior.[36] Immunity is particularly interesting in light of today's COVID-19 stress, and also the rise of autoimmune disorders worldwide: roughly 4 percent of the world's population is affected by one of more than eighty different autoimmune diseases, one of the most common of which is type 1 diabetes.[37]

The immune system is "educated" and matured by commensal bacteria, especially bacteria in the gut. An altered gut microbiota influenced by these different factors can directly affect immune cells in the gut—and indirectly affect immune cells through microbial products like metabolites, which can affect insulin resistance. Therefore, the human microbiome has the potential to affect the patho-

The Gut-Friendly Diet

Keeping your gut microbiome healthy may be the secret to weight control.[38] There are microbiome-boosting supplements, but the real trick is a phased approach of supplements and diet to repopulate your gut with healthy bacteria. One of the pioneers of the microbiome diet is Dr. Raphael Kellman, a board-certified physician specializing in gut health. He outlines the three phases you need to go through in his book *The Microbiome Diet* (Hachette Books, July 2014). In the first twenty-one days (Phase 1) he suggests you adopt an organic, plant-based diet with prebiotic-rich foods, such as asparagus, garlic, onion, and leeks, along with fermented foods that are rich in probiotics: sauerkraut, kimchi, kefir, and yogurt. You should avoid: all grains, eggs, most legumes, dairy, packaged or fried foods, sugars, fish, meat, and alcohol.[39]

genesis of immune diseases, like type 1 diabetes or autoimmune diseases such as Lupus, Crohn's disease, rheumatoid arthritis, inflammatory bowel diseases, and coeliac disease.[40]

Microbes May— or May Not—Like Where You Live

But let's get back to the discovery of how geographical location—where you were born or where you live—the trajectory of incidences found in autoimmune disease, and the potential disruption from changes in the gut microbiota are of importance. Mapped from the review of several independent studies, the results showed that autoimmune disease incidence in Europe follows a North-South trajectory (with Sardinia, Italy, as one notable exception), with higher instances found in Nordic countries, such as Finland, Sweden, and Norway. Other gut microbiota studies delved into the effect of extreme climate conditions (i.e., polar expeditions) and even birth month/place, illustrating how climate, particularly sunlight exposure, affects microbiota composition and immune dysregulation. And on that sunlight exposure, researchers also found the factors of circadian rhythm disruption and vitamin D deficiency to be prominent in northern populations, linking to immune dysregulation through shifts in gut microbiota—which, as we see now, connects to autoimmune diseases like type 1 diabetes.[41]

Gut Virome

Bacteria aren't the only microorganisms squatting in your gut. The city inside your gastrointestinal tract consists of bacteria, archaea, fungi, and viruses. More recently, researchers have begun to investigate the role of the gut virome as it relates to both states of health and disease. Understanding the character development of the intestinal virome from birth to the onset of autoimmunity has the potential to be an important component to understanding the cause of type 1 diabetes.[42]

Live Long on the Sardinian Diet

Sardinia, Italy, is one of the original "blue zones" identified as a region with extreme longevity. In 2004, researchers found that the lifelong residents of Sardinia have a rare genetic marker for exceptional longevity called the M26. They found that because of Sardinia's geographic isolation, they still hunt, fish, and harvest their own food. Key to their healthy long life is their diet: a lean, mainly plant-based diet with very little meat. They mainly consume whole-grain bread, beans, garden vegetables, and fruits. They also take a walk every day and drink a glass (or two) of red wine daily.[43] Do you think their gut is happy? If the map of autoimmune diseases in Europe called Sardinia "the exception," there may be something there.

Gut and Marriage

Here's something to ponder: your long-standing relationship (marriage to some) can influence your gut microbiota. Analysis of the gut microbiome of spouses and siblings revealed that spouses have more similar microbiota and more bacterial taxa in common than with their siblings. Turns out marriage does have its microbial advantages: married individuals harbor microbial communities of greater diversity and richness relative to those living alone, with the greatest diversity among couples reporting close relationships.

—*Nature*[44]

Next Time Someone Asks Your Age, Tell Them to Speak to Your Microbiome

It seems that the microbes on our skin and in our guts can reveal our chronological age with near-perfect accuracy. A team of researchers from the University of California, San Diego, working with the IBM Thomas J. Watson Researcher Center, Yorktown Heights, New York, have found that your skin microbes can be used to predict your chronological age within about four years.

Drawing from nearly nine thousand fecal, saliva, and skin swab samples from people in the United States, the United Kingdom, Tanzania, and China collected as parts of other microbiome projects, the researchers built a computational model that compared each person's age to the types and number of microbes they had. The team then tested the data by working backward to see how well microbiomes alone predicted age. Skin microbiomes were the most accurate, whereas stool microbes, reflective of the gut, came in second, predicting real age within about eleven and a half years.[45]

In a similar study, and with the help of artificial intelligence and 3,600 samples of gut bacteria from 1,165 healthy individuals around the world, longevity researcher Alex Zhavoronkov and colleagues at Insilico Medicine (a Rockville, Maryland–based artificial intelligence startup), found that their program was able to accurately predict someone's age within four years—based just on their gut microbiome. Like the methodology used by the University of California, San Diego, and the IBM team, Insilico trained their computer program on different species of bacteria from 90 percent of the samples, correlating them to the ages of the people the samples had come from. The researchers then asked the algorithm to predict the ages of the people who provided the remaining 10 percent of the samples. An interesting insight from their research is that thirty-nine out of the ninety-five species of bacteria studied were observed to be the most important in predicting age.[46]

While these studies are just cracking the surface of how our microbiome reflects our age, we still know little of how the microbiome changes over time, or even what is considered a "normal" microbiome. As the body of microbiome data expands with more samples from clinical studies and citizen scientists like you (and may we all thank you for your contribution!), we will realize amazing new insights that could help develop noninvasive microbiome-based tests to determine signs of accelerated or delayed aging, and design microbially based interventions to ameliorate the aging process.

Personalized Medicine 2.0

Today, there are still many diseases that we don't yet understand at a molecular level—from Alzheimer's and Parkinson's to food allergies—and the microbiome could be key in their possible treatments, including drug development.

With further understanding of how the specific bacteria in the microbiome help boost the immune system, doctors could one day prescribe "medicine" that is a personalized mix of living microbes, and not chemicals from a lab. Perhaps a microbial swab will be as routine as taking a blood test. Yearly exams may include fecal samples to profile your gut microbiome, tracking changes in your microbiome to predict or prevent your risk of disease.[47]

As a society we have become addicted to antibiotics. They are great drugs for serious illnesses, but are being used more and more to treat ever milder conditions, in which their net positive effects are marginal. We clearly have to restore our lost microbes. Meanwhile, my colleagues and I have been working to create a microbiota vault, where we can preserve our ancestral microbes for future generations, before many important ones become extinct. Thankfully what we've learned in recent years about the microbiome may enable us to live more collaboratively with bacteria. In the not so distant future, for example, pediatricians may examine both babies and their diapers to determine whether that infant has an ideal microbiota, based on their genes and other markers. If not, they will be able to administer the "missing microbes" to optimize the baby's health trajectory.

—Martin J. Blaser, from "Human Health Is in the Hands of Bacteria," *Time*[48]

How Do You Feel? The "Psychobiome" and Your Gut Bacteria Want to Know

That "gut feeling" may be related to what's going on with your microbiome. As we learned, your gut microbiome produces and regulates neurotransmitters, and shares them with the brain, linking the gut to the mind. Further research into this connection, and the bacterial mix that is critical to it, could lead to the development of new mental health therapies. A growing number of researchers see a promising alternative in microbe-based treatments, or "psychobiotics," a term coined by neuropharmacologist John Cryan and psychiatrist Ted Dinan, both at University College Cork. The term psychobiotic is a targeted intervention of the microbiome for brain health. And while a probiotic for mental health may still be in the far future, when it arrives, it will become routine for doctors to keep an eye on the makeup of patients' microbiomes not only for physical but for mental health.

Holobiome, a startup based in Cambridge, Massachusetts, has created one of the world's largest collections of human gut microbes in hopes that by isolating and culturing bacteria they may eventually produce new treatments for depression and other disorders of the brain and nervous system. The targeted ailments include depression and insomnia, as well as constipation, and visceral pain like that typical of irritable bowel syndrome—conditions that may have neurological as well as intestinal components.

The researchers at Holobiome also discovered so-called GABA-producing gut microbes play a key role in mood disorders. GABA (gamma-aminobutyric acid) is an amino acid and neurotransmitter that inhibits neural activity in the brain. GABA is what helps us with sleep, reduces mental and physical stress, lowers anxiety, and creates a calmness of mood. Its imbalance has been linked to depression and other mental health problems.[49] Holobiome has identified and ranked thirty promising GABA-producing bacteria, including the ones they are testing.

Probiotics That Are Psychobiotics

Researchers are just beginning to identify which probiotics may have mind-altering effects, for decreasing depression and anxiety. Some include:

Probiotic pills that consisted of *Lactobacillus acidophilus*, *Lactobacillus casei*, and *Bifidobacterium bifidum* (2 billion CFUs each) significantly decreased total scores on the Beck Depression Inventory, a widely used test to measure the severity of depression, after eight weeks of usage.[50]

A probiotic containing *Lactobacillus helveticus* R0052 and *Bifidobacterium longum* R0175 (Probio'Stick®) alleviated psychological distress, depression, and anxiety, and improved problem solving when taken for thirty days.[51]

Until Your Bespoke Psychobiotic Arrives . . . Improve Your Personality with Microbes!

By now we should understand the important link between the gut microbiome and our brain, behavior, and overall health (the BIG axis)—but what about our personality traits? That's one area that interests Katerina Johnson, PhD, Experimental Psychologist at Oxford University, who set out to examine if there was a connection between the composition of the gut bacteria and personality traits such as sociability and neuroticism. And the result could make you and your Instagram followers happy: people with robust social networks were more likely to have a more diverse composition of gut bacteria.

The study collected fecal samples from 655 adults, 71 percent of whom were female and 29 percent male, with an average age of forty-two. The study also involved the participants answering a comprehensive questionnaire that inquired about their behavior, health, lifestyle, and sociodemographic factors. Johnson carried out a set of statistical analysis to help determine the relationship between the composition of the gut bacteria and behavioral traits such as sociability and neuroticism.

This is the first study to find a link between sociability and microbiome diversity in humans and follows on similar findings in primates, which have shown that social interactions can promote gut microbiome diversity. The analysis also revealed that lower microbial diversity was associated with higher levels of stress and anxiety.[52] Seriously, it's time to get more social. Your gut will thank you.

Eat Your Way to a More Social Microbiome

So what's on the menu? People who ate more foods with naturally occurring probiotics or prebiotics had significantly lower levels of anxiety, stress, and neuroticism. However, Johnson didn't find the same correlation with probiotics or prebiotics in supplement form. Natural sources of probiotics include fermented cheese, sauerkraut, kimchi, and natural sources of prebiotics include bananas, legumes, whole grains, asparagus, onions, and leeks.

Another intriguing finding was that people who had been fed formula as infants had a less diverse gut microbiome. This is the first time this has been investigated in adults, and the results suggest that infant nutrition may have long-term consequences for gut health.[53]

Carpe Diem . . . or *Mycobacterium Vaccae!*

Bacteria Is for Learning

We learned in chapter 1 that bacteria are smarter than we think. Now, studies are showing the potential of how, when we spend more time in nature, we may be absorbing more of one of the smarter bacteria—*mycobacterium vaccae*, a natural soil bacterium to be exact. When we walk on grass or spend time in wooded areas, we humans likely touch or breathe in smarty *M. vaccae* bacteria. Previous research studies demonstrated that when *M. vaccae* bacteria were injected into mice, it stimulated growth of neurons that encouraged an increase in levels of serotonin, and a decrease in anxiety. Since serotonin also plays a role in learning, Dorothy Matthews of the Sage Colleges in Troy, New York, and her colleague Susan Jenks wanted to see if live *M. vaccae* could improve learning in mice.

The researchers modeled a "race" to navigate a maze between two groups of mice: one fed live *M. vaccae* bacteria and the other were not fed any bacteria. What they found is that mice that were fed live *M. vaccae* navigated the maze twice as fast—and with less notable anxiety than the control mice.

While the research suggests that *M. vaccae* may play a role in both anxiety and learning in mammals, Matthews speculates that perhaps "creating learning environments in schools that include time in the outdoors where *M. vaccae* is present may decrease anxiety and improve the ability to learn new tasks."[54] Interestingly, during the peak of COVID-19, some areas were moving class-rooms outdoors—where students would be likely to get a dose of *M. vaccae* along with their times tables. Wonder if they noticed any change?

Fall in Love with Future Wellness

Probiotic Bacteria and the Love Hormone

Dr. Susan Erdman at Massachusetts Institute of Technology, Boston, is exploring how parent-infant bonding after birth not only enhances a baby's physical health but can also impact the potential for a virtuous life, with effects extending to future generations. Dr. Erdman's premise and goal is to establish a generational paradigm of increased human virtue, based upon novel findings involving probiotic bacteria and the "love hormone" oxytocin in mouse models.

According to the published grant, early life events and the brain hormone oxytocin are central in human pair bonding. The study proposes a microbial restoration strategy to stimulate natural oxytocin secretion, based on a phenomenon discovered in mouse models.

Creative collaborations between MIT and Harvard-MGH will use probiotics originating from healthy human breast milk, namely *Lactobacillus reuteri* 6475, to stimulate oxytocin and beneficial parent-infant bonds. The team hypothesizes that oral supplements with *L. reuteri* will boost oxytocin levels and favorable behaviors including empathy, altruism, and spirituality in mothers and their infants. Actionable outputs will inspire public health change, with vast potential to improve human existence.[55]

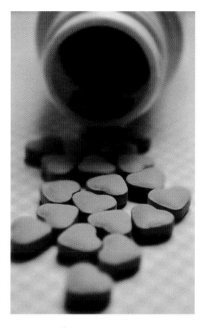

What if our gut bacteria really do guide us to act with a sense of purpose? As it turns out . . . the molecule behind a meaningful life may well be hypothalamic mastermind oxytocin. Best recognized as the "love hormone," oxytocin also has roles in childbirth, parenting, and spirituality. Furthermore, oxytocin imparts an otherworldly sense of connectedness with the universe. This transcendental aspect of oxytocin raises the intriguing possibility of a deeply enriched human experience on the other side of microbes. If our ultimate goal is a physically healthy *and* purposeful life, then our microbial passengers and oxytocin are important partners in our journey.

—Susan Erdman, principal research scientist and assistant director in the Division of Comparative Medicine at MIT[56]

Speaking of Love . . . Fido Acts as a Source of Healthy Bacteria?

Does a canine member of your family affect the microbiome of its caregiver? Well, YES! Bacteria from Fido's fur and paws is easily transferred to the skin of its human family living in the same home.

A study at the University of Colorado in 2013 showed that simply owning a dog had an impact on the sharing of microbes between one person and another living in the same place. Even more interesting is that cohabiting couples who owned dogs had more bacteria in common with each other than couples who didn't have dogs. The sharing process of their microbiome was in the pets—when one partner strokes a dog, they leave behind bacteria, and when the other partner gives their dog some loves, they pick up their partner's bacteria.

There are many health benefits to owning a pet, and dogs get a high rating: people who own dogs are happier, less stressed, and even less likely to die of heart disease. But the University of Colorado research begs the ques-

tion: Could a dog, perhaps, be a type of probiotic for us humans? Could your dog help spread beneficial bacteria?

Two studies looked at the health benefits of living with a dog, one in children and the other in older people. University of California, San Francisco, scientists suggested that living with a dog starting at infancy may lower a child's risk of developing asthma and allergies, largely as a result of exposure to what they call "dog-associated house-dust" (also the proverbial "dust bunnies"). The researchers hypothesized in 2013 that babies and small children need to be exposed to harmless bacteria in order to "train" their developing immune systems, and a dog is a perfect vehicle for that kind of exposure.[57]

A more recent study explored whether dogs could directly improve the health of older individuals. The researchers in Arizona adopted unwanted dogs from the Humane Society, then gave the dogs to people over fifty who either never owned a dog

"Fitting In" with the Rest of Life in Our Social Distance Bubble

There's a whole world unfolding in our homes, one that we will never see: the life of our home's microbiome. This collaborative world of bacteria, fungi, viruses, and other microbes is on every surface and even in the air we breathe.

In chapter 3 we will look at rewilding our urban biome, and how the quarantines and stay-at-home orders of 2020, besides wreaking havoc on our social relationships, have transformed our homes' microbial ecosystems. For now, we want to focus on the challenge of balancing the good, the bad, and the ugly in our microbial worlds—in other words, how should we keep our house free from any dangerous microbes that might make us sick, but still keep the good bugs?

Researchers have found that simple changes such as replacing your kitchen sponge weekly and regularly washing your kitchen

or hadn't had one for a while. They're monitoring the physical and mental health of both owner and dog to see if the good bacteria from the dogs is transferred to their new owners, along with other health-boosting benefits.[58] So next time you are feeling a little down, give your dog a big hug and share those good bacteria!

towels—two things that are regularly damp and that touch both you and your dishes often—is a good way to prevent the growth of microbes that might cause human disease. Generally speaking, keep surfaces and fabrics as dry as possible to avoid the moist, damp environments in which bacteria thrive.

If you are still burning with curiosity about what might be living in your home, *Popular Science* suggests checking out *Community of Microbes*, an app and website created by an artist and a microbiologist that tells you more about the different kinds of microbes living in your environment.[59] And be a good roommate: stay away from too much antibacterial hand soap, which wipes out your skin microbiome (. . . post COVID-19, Dr. Fauci will forgive you!).

Community of Microbes

Using augmented reality, you can go on a virtual microbial hunt around your home with the Community of Microbes app. Look to your floors: they may be dry, but they are unforgiving deserts for the microbes that live in our homes. Explore the tiny balls of textile fibers, bits of insect carcasses, and giant boulders of pollen granules that dwell there. Many of these microbes have managed to find ways of surviving the dryness of the floor by going into a state of sleeplike dormancy, protecting themselves from dehydration. They wait like seeds in the desert for rain, perhaps hiding in a pile of wet laundry, a forgotten wet crumb of food, or even found lounging on a very, very humid day. With a bit of moisture these infinitely small roommates awaken from their quiet slumber around us. The home environment can contain hundreds to thousands of different species of these microbes.

—Community of Microbes[60]

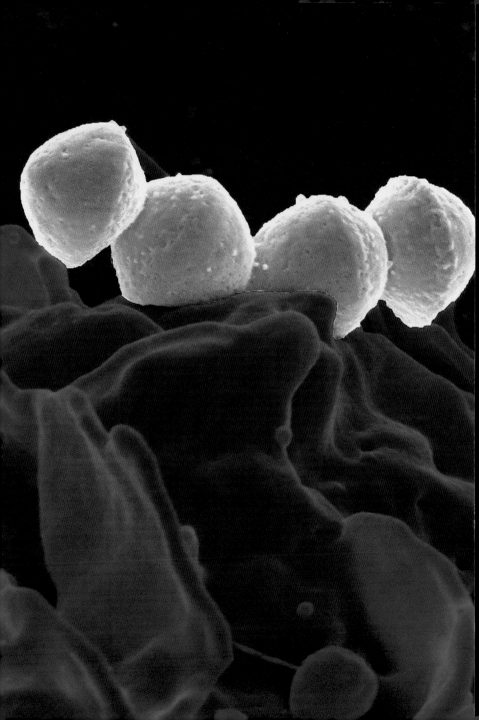

Rewilding Our Urban Biome

Everywhere and the Kitchen Sink

We're learning that our microbial fingerprint is not ours alone—it's born from the complexity and interconnectivity between our microbiome and the world. Your personal microbial gardens need to be pruned and replanted to keep its community healthy. How do we do that? By venturing out in the "wild"—everyday life where we introduce our microbiome to what's out there (besides us). Our microbiomes need to mingle, cross-pollinate, and get caught in the clouds of other microbiomes living on everything that surrounds us. Time to spread your microbial dust around!

You Are Not Alone, Ever

You are surrounded in an invisible cloud everywhere you go. Each person sheds a distinct combination of microbes and, like the data clouds that contain your digital fingerprints (tracking and storing your information), these microbial "clouds" can be used to reveal a person's gender, age, link people to geographic locations, and much more.[1]

Everyone's unique cloud contains millions of their own bacteria that never leave their side. This ever-present colony is a by-product of our microbiome, which researchers have found emits as many as one million biological particles that escape our body and can leave a microbial fingerprint on our surroundings (what microbiologists call the "built environment"). For example, when we come in direct

contact with surfaces we touch or rub against, or when the tiny pieces of skin or hair we shed get moved about, they kick up a dust of microbes that escape. The bio-composition of our odors, from our breath, skin, and hair, also leaves traces. We can often track microbial communities in the built environment back to an individual person, based on their direct contact with an object or surface, from a school desk to a mobile phone.

It was a team of researchers from the University of Oregon that first exposed the microbial cloud when they found that the human microbiome emits trace biological particles. In one experiment, the researchers placed volunteers in a climate chamber—an enclosure that looks like a vault used to test the effects of environmental conditions on specimens. The team took samples of the air inside the occupied chamber and when they compared its microbial makeup to that of a sterile climate chamber, the results confirmed that the occupied chamber had a microbially distinct print. Moreover, they were able to detect which volunteer was in the space. The study demonstrated for the first time that each of us release our own personalized microbial cloud.[2]

Sounds strange, right? A cloud of microbes swirling around us and interacting with the clouds of everyone and everything in our surroundings? In trying to imagine your own cloud, it may be helpful to call on the same metaphor Stanford researchers used in their own observations: They dubbed the intermixing of microbial clouds a "Pig-Pen" after the beloved *Peanuts* character. Unsure how Charles Schulz would describe the composition of his character's cloud, the Stanford University team described our "pig-pen" cloud as one part environmental and shared with immediate neighbors; and one part more personal, consisting of human and pet-centric bacteria, fungi, parasites, and protozoa.[3]

The uniqueness of our microbial clouds is especially interesting to researchers studying epigenetics. We know now that the effects of your environmental exposures—lifestyle, diet, stress, the effects of drugs, or instances of other diseases—can trigger a particular gene to turn on or off. It seems that the microbes that exist in our clouds and our surroundings are now being considered part of that pool of environmental epigenetic stressors,[4] adding another crucial piece to our personal health puzzle. An especially large body of emerging research is looking at the epigenome and microbiome as part of a larger portrait: the exposome.

You, the World, and Your Exposome

The exposome can be defined as the measure of all the exposures of an individual in a lifetime and how those exposures relate to health. The term "exposome" originated from a cancer researcher named Christopher Wild, a result of his concerns for unexplained diseases that led to cancer. This was in 2005, when the science community was sequencing the human genome, and its hope for the holy grail of disease origins had overshadowed the importance of environmental exposures on our health.[5]

These exposures begin before birth and include factors that are environmental and occupational.[6] (Think everything from childhood secondhand smoke, to

soil contaminants that hang on to fruits and vegetables, to the dust in your home, to Fido's dander.) But it hasn't been until recently that researchers have considered an individual's daily exposure to airborne bacteria, viruses, protozoa, fungi, and chemicals as part of the exposome.

In short, your exposome lives in a world of "others." Other people, other animals, other environments, other forms of transportation. The challenge of understanding the impact of living in diverse microbial clouds is how we can detect and monitor all these different clouds we breathe in every moment. One possible solution came from a team of researchers at the Department of Genetics of Stanford University School of Medicine in California. They built a wearable device that looked like an external flash drive, which continually sampled the air that their fifteen participants were exposed to when they were at home, at work, or on the

We Inhale Earthy "Joy"

In their exposome tracking study, a team of geneticists from Stanford University School of Medicine discovered that many of the samples they gathered contained the earthy chemical compound known as geosmin. Geosmin causes that earthy smell that we humans like to take a deep breath of when we are out in nature, and most likely gives us a moment of joy. To the researchers, this was an example of things we are exposed to that don't necessarily harm us.

—*Scientific American*[7]

road, collecting the various biological and chemical compounds.

After three months of use, the team analyzed the filters using DNA sequencing and found that the fifteen participants had been exposed to 2,560 biological species. The team also used mass spectrometry analysis to see the chemical signals in the mix and found almost all the samples contained diethylene glycol, which is used in products from brake fluid to skin cream, and some traces of the insect repellant DEET.[8]

The micro-world is showing clear signals of the large and slow ecological crisis we're in. We've already lost a third of microbial diversity. As we are headed towards dramatic ecological and microbial collapse, the Architectural Exposome aims to design tools that push architects to engage with ecosystems by allowing for direct interaction with the microbial.

—Ioana Man, multidisciplinary designer working among architecture, set design, and critical practice[9]

Let's Get Rewilded . . .

Promoting connections to the microbes many of us currently shun should be a key part of any post-pandemic recovery strategy. We must protect and promote the invisible biodiversity that is vital to our personal and planetary health.

—Jake M. Robinson, PhD, researcher, Department of Landscape, University of Sheffield, via *The Conversation*[10]

Rewilders Unite!

Like we discussed in chapter 1, humans and our microbial dwellers have a collective genome and a life history together. So much so that microbiologists call the microbes found in soil environments, places like grass or the woods, our "old friend" microbes. They hypothesize that these microbes evolved alongside humans, settling in with each migration through the centuries. These soil microbes populated us as we populated the planet; we are truly a master host of symbiotic microbiota.

"Old friend" microbes play a

major role in educating our immune system to help fight infectious diseases like COVID-19. Our immune defense is impossible without support from diverse microbiomes, and just as microbes have important roles in ecosystems–helping plants grow, recycling soil nutrients, etc.–they also provide our bodies with nutrients and health-sustaining chemicals that boost our immune system and promote good physical and mental health.

Biodiversity is the vital variety of organisms, plants, and soil ecosystems that make up our natural resources, our lifeline to food security, and the foundation of our health and that of the planet. The activities of humans are creating a severe biodiversity loss. Industrial urbanization is a major factor, disrupting the symbiosis between microbiota and their hosts. Our industrialized urban habitats are low in microbial biodiversity, upsetting the habitats in which microbes colonize.

Without a healthy environment, the social ways microbes fend for their colony or send chemical signals to defend invasions are weakened. That means the whole ecosystem becomes weak–

Seeking Interdependency

There are interdependencies between biodiversity, holobionts, and public health that lead us to argue that human health outcomes could be improved by increasing contact with biodiversity in an urban context.
—*Frontiers in Microbiology*[11]

Roughly one million animal and plant species are now threatened with extinction, many within decades, more than ever before in human history. Seventy-five percent of the terrestrial environment has been "severely altered" to date by human actions.
—UN Report[12]

including the one we carry with us. These habitat factors, such as the concrete jungles we have erected since the turn of the nineteenth century, industrialized diets, drugs, and antibiotics, are now associated with the epidemic of noncommunicable diseases (heart disease, cancer, chronic respiratory diseases, and diabetes) in urban societies. What's at stake is the restoration of urban microbial biodiversity. The solution? Micro-ecological processes that encourage microbiome rewilding to strengthen the health of our microbial partners and us, and ultimately aid in preventing urban noncommunicable diseases.[13]

Which brings us to the conundrum of germs vs. sterilization. Of course, well before COVID-19 a number of scientists had warned us of the rapid loss of our microbial diversity—we have yet to realize the immense impact of it, especially in our urban ecosystem. Unfortunately, COVID-19 has caused us to not embrace microbial diversity, but rather, in a desperate if ill-advised attempt to insulate ourselves from the virus, encouraged us to readopt a global culture of germapho-bia. Once the pandemic hit, people around the world panic-purchased antibacterial products in bulk, rubbing antibacterial sanitizers on their hands incessantly, and wiping all the surfaces in their homes, cars, and everywhere they could with sterilizing solutions. Stay-at-home orders and closed playgrounds prevented children from going outside to play, which we know is essential to their immune defense. (By the by, if you live in a city of mostly concrete, you are probably missing out—pandemic or no pandemic.)

Unfortunately, that germophobic impulse to kill all the bacteria around you could have serious consequences, such as the creation of resistant superbugs. We need microbes—bacteria, fungi, and, yes, even viruses—to maintain healthy ecosystems. Humans are a part of these ecosystems, and our health suffers when our microbial cloud is depleted. We need to reset the clock to pre-COVID-19 days when the US Food and Drug Administration (FDA) advised people to stop using antibacterial soap (which is no more effective at preventing illness than regular soap and may negatively impact health).[14] Before the pandemic, some health experts urged people to cut back even on alcohol-based hand sanitizer in part because some bacteria have become more alcohol-tolerant.

It's time to change the narrative and embrace our inner Pig-Pen. So, the next time you see a patch of dirt, mix it up with your hands, and reacquaint yourself with your old microbial friends.

Outside-In: Our Microscopic Roommates

Just like we need to get serious about the urban microbiome, we also need to get serious about our home biome. The home biome is the community of microorganisms (bacteria, fungi, and viruses) that inhabit your home or apartment, and everything that lives with you. Humans spend most of their time indoors—Americans spend as much as 90 percent of their time inside (90 percent!!)[15]—and Earth's fastest growing biome is the indoor environment. There is evidence of interaction between the chemical components of our

Don't buy into the hype of antibacterial soaps. According to the FDA, antibacterial soaps aren't more effective than plain soap and water for killing disease-causing germs under most circumstances in the home or in public places. Additionally, some scientists caution that the long use of antibacterial soaps over time may contribute to the development of germs that are antibiotic resistant. The best recourse in your clean fight against germs is thorough handwashing—scrub in between your fingers, the palms, and tops of your hands—for at least twenty seconds to stop the spread of infection. Need a timer? The CDC suggests humming the "happy birthday" song.
—US Food & Drug Administration[16]

house dust and the biological community of the home biome. We share our space with massive amounts of microscopic life that, for the most part, are our friends.

A 2014 study conducted by researchers from the University of Chicago and the US Department of Energy's Argonne National Laboratory looked at the microbiome of an entire family and their pets in the home environment, following them over a course of six weeks. The Home Microbiome Project involved seven families (eighteen people), three dogs, and one cat. The samples were gathered two ways: The participants had to swab their hands, feet, and noses daily (dogs and cats had some help from their human parents) to collect a sample of the microbial populations living in and on them. Additionally, the surfaces of the house were wiped for samples, including doorknobs, light switches, countertops, and floors. The study found that people directly impact their microbiome, and that of their home. When three of the families in the study moved, it took less than a day for the new house to have a similar microbial profile as their former home. The research also suggests that when a person and their microbes leave the house, the home's microbial community shifts. And not to forget our old friend soil microbes, the results also showed more plant and soil bacteria in the houses with indoor-outdoor dogs or cats. This research gives promise to the fact that one day we could theoretically predict whether a person has lived in this location, and how recently—a biological form of contact tracing, or to help with missing persons.[17]

Understanding where a person lives sheds further insight on their environmental exposure and the hazards and risks that they will encounter. New tools using artificial intelligence and crowd-sourced data, such as the Map My Environment global initiative,[18] can help create a complete understanding of a person's health as well as that of their community. Certain diseases and cancers have been linked to environmental factors. The more we can learn about how our environmental exposures interact with our microbiome and genetics, the more proactive we can be.

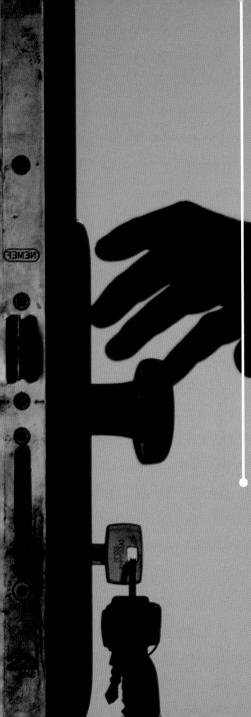

Do You Know What's in Your Home?

The average American household has thousands of different fungal and bacterial organisms. Many are harmless, some help protect you, and others can harm you. Homebiotic helps you discover your home biome, and ways to optimize it for better health with the use of environmental probiotics, air purifiers, or other technologies.

According to Homebiotic, their Home Biome test covers all the major types of mold that ERMI tests do, and can also identify bacterial organisms to provide a comprehensive picture of what's living with you in your home environment. Samples are tested by the laboratory at the Center for Medical Mycology at the Cleveland Medical Center, using advanced quantitative polymerase chain reaction (qPCR) and NGS DNA sequencing technology.

The easy-to-do test comes in a kit with sterile collection swabs, a pre-addressed return envelope, and a registration ID for your samples.
—Homebiotic[19]

Whatever else is happening outside our front door, at least we can have a powerful and positive influence on our home biome through the products we buy, the foods we eat, the chemicals we clean with, and the outdoor samples that we bring home with us each day.

The biggest impact we can possibly have on improving the health of ourselves and our family is to start treating the home environment as you would treat your body. Cleaning products spiked with antimicrobials are not actually doing much to boost the sanitizing properties of the objects you're cleaning. As we discussed, antibacterial soap is a good example—it really comes down to how thoroughly you clean, instead of the extra chemicals. While microbes can develop resistance to drugs that target specific parts of their life cycle or anatomy, the same isn't true of soap. Avoiding soap-based destruction would require a pathogen to grow an iron cage around its cell, which is impossible.[20]

There is no reason to go crazy on the cleaning. The beneficial bacteria that help our bodies function on a daily basis could suffer from overexposure to cleaning compounds. In recent years, scientists have come to realize that grime, grit, and germs are actually pretty important for training our immune system during childhood. And while you certainly don't want mold or excessive grime in your home, you can save that Comet for another time.

We spend most of our waking life indoors, and it's time to nurture the biodiversity inside our walls. We can no longer think of the "outdoors" or "nature" as an ecosystem separate from the "indoors," because there is one common carrier that cross-pollinates both nature and the built environment: us. We are the perpetrator and the gardener of the interconnected web of life that biodiversity, both in the wild and in the urban field, makes possible. Restoring natural habitats inside and outside of your residence can help increase biodiversity and the health of our homes.

Home Sweet Phyllosphere

The Green Urban Microbial Thumb

If you're one of the lucky ones that manage to keep your house plants alive, you are helping your home biome (and possibly your own). Plants have a way of controlling microbial diversity within their leaves to promote their health. Scientists at Michigan State University in East Lansing have shown how plant genes select which microbes they allow to live inside their leaves in order to stay healthy. Researchers mostly understand what happens with the microbial communities in the soil that help plants grow, but one of their studies was the first to expose the relationship between the plant and the crowd of microbes assembling in the aboveground portions of plants (known as the "phyllosphere").

What the researchers learned is that the plant's genes act like gatekeepers for the possible millions of microbes bombarding the plants in our homes and in our neighborhoods. The plant's genes are the selection committee, helping choose which microbes can stay, and who needs to go. The plant's decision process involves two genetic networks: one that runs the plant immune system, and one that controls hydration levels inside leaves. Both networks appear to work together to select which microbes can coexist with the plant. It appears that organisms, from plants to animals, may share a similar strategy to control their microbiomes.[21]

The potting soil or mulch you use for indoor plants or landscaping has a healthy dose of diverse soil bacteria that is good for your immune system. Researchers are studying how *Mycobacterium vaccae*, harmless bacteria dwelling in soil, communicate with our own cells, based on mouse studies where heat-killed *M. vaccae* had immune-modulating and mood-altering properties.[22] The next time you plant—whether in your front or back lawn, or potted or raised gardens on your balcony or doorstep—be mindful that the plants are not just a thing of beauty, but a potential contributor to the health of your urban and home microbiome. City garden projects are one way to help restore diverse plant communities in urban areas. More plant life in our city sidewalks may provide human health benefits by diversifying the overall environmental microbiota. While there is still future work to do on designing, building, and managing urban green spaces, actively contributing to the urban garden outside your door puts your own microbial green handprint on a healthy city.[23]

Microbes Built This City

You could say microorganisms like bacteria, viruses, and other pathogens had an influence on how modern cities have evolved, especially after historical epidemics such as bubonic plague and cholera, or viruses like influenza. These deadly bacteria and viruses forced former generations to rethink the health of their homes and the city they dwell in. Over time, urban planners and city health officials around the world have implemented better sanitation systems, redesigned airways, and improved air circulation in public buildings, in an effort to deflect a contagion. London, for example, changed the construction of newer buildings following the Great Plague (1665) and the Great Fire (1666), from timber to the beautiful brick buildings we see today, as brick is less flammable, and also harder for rats to burrow into (the animals that spread the plague).[24]

Infectious disease had an influence on how New York City was designed, as presented at the compelling "Germ City: Microbes and the Metropolis" exhibition at the Museum of the City of New York (MCNY). On display from October 2018 to April 2019, the exhibit explored how the relationship between humans and microbes has "had a profound influence on human history—especially in cities, the crossroads of the movements of people, goods, and germs."[25] It gave a historical and modern look at the influence of microbes in the design of New York City, from housing and sanitation to streets and public spaces. Co-organized by the New York Academy of Medicine and Wellcome, a biomedical research charity, the exhibit portrayed how infectious disease influenced the city physically, culturally, and sociologically.

Throughout the history of the evolution of cities, two underlying forces were at play, both unknowingly working together: people and microbes. In the 1800s, epidemics of diseases like cholera, measles, scarlet fever, and tuberculosis swept the streets of London, New York, and other rising global hubs. One of the reasons was poor sanitation, since local water supplies were frequently contaminated. Once city officials in New York realized that the city's dirty water was a source of breeding disease, they developed an aqueduct system that brought fresh water from north of the city into New York City's water systems—systems the city still relies on to this day.

Another microbial challenge of the 1800s was the transportation of the day—horses. London and New York were faced with millions of pounds of manure that would pile up on, you guessed it, the streets. The mounds of manure were a breeding ground for microbes, and the flies that fed on the manure were the deviants that spread diseases like typhoid, cholera, salmonella, and dysentery. But this challenging cesspool led to the development of modern sewer systems, and New York owes thanks to a former Civil War colonel named George Waring for developing the intricate NYC system that exists today.[26]

As global society marched into the twentieth century and the global industrial revolution, cities became a place of hope for immigrants looking for work, and cheap housing was in demand. In New York City, immigrants in the early 1900s were living in overcrowded tenements, where the conditions were extremely challenging for anyone's health, with poor lighting, air circulation or ventilation, and the lack of indoor plumbing. In such cramped quarters, disease spread quickly, and the most notorious of pathogens was "The White Plague" or tuberculosis.[27]

We still have much work to do in improving living conditions in cities today the world over. City planners are rethinking public and private spaces, parks, and community gardens to provide more access to fresh air and open space, and planting an urban biodiversity of plants, birds, insects, and small animals—and yes, the industrious, happy microbes that inhabit it.

Wilding Our Buildings

In his text *On Architecture*, Roman architect and engineer Vitruvius suggested that architecture is an imitation of nature. Architects and material designers have been adopting the principles of biomimicry—emulating strategies from nature to sustain and thrive—for developing next generation energy-efficient materials and methods of construction that are adaptable to environmental conditions. Today's urban engineers are working on ways to make buildings require less energy and become self-sustaining, like capturing rainwater to filter into the water system, and they are looking to biological systems for inspiration.

One biological system that has bioengineers excited about future development of architecture is the bacterial biofilm. You'll remember that a biofilm is the complex structures different colonies of microorganisms create by attaching to and forming a community on inert (e.g., rocks, glass, plastic) or organic (e.g., skin, cuticle, mucosa) surfaces.[28] The properties of the surface such as the electrical charge, dampness, or roughness, determine how bacteria initially attach. A common characteristic of all biofilms is the production of an extracellular matrix (or ECM), which is a three-dimensional network composed of different organic substances that provide structural and bio-

chemical support to surrounding cells.[29] Biofilms offer bacteria several ecological and physiological advantages, one being the physical barrier against host defenses during infection, and the other being the protection from antimicrobial agents (such as disinfectants and antibiotics).[30]

Researchers from the School of Architecture, Planning and Landscape, at Newcastle University, Newcastle upon Tyne, United Kingdom, are proposing that bacterial biofilms, which are necessary in soil biomediation (which in today's sustainability lingo means "regeneration," or restoration of soil back to a healthy state) have a potential use in architecture and construction. They suggest that biofilm communities enhanced by synthetic circuits may be used to construct buildings and to sequester carbon dioxide in the process.[31]

Since soil microbes are so good at carbon sequestration, rewilding biodiversity is being praised as a climate-action strategy—a biological holy grail that can be put to work in urban and rural areas. From the diverse microbial populations living in micro-caves underground, to the airborne clouds of microorganisms and everything in-between, the loss of healthy, commensal, and social microbial populations in the so-called sterilization of cement landscapes is not good for humans or the planet. Once again, architects and progressive urban planners are taking notice, and have begun rethinking cities as multispecies cohabitation and finding ways for us to start living more closely with animals, possibly creating an interdependent urban ecology.

More urban designers are bringing back gardens to the city in hopes to reintroduce some form of circular agriculture, where the garden is the local source of vegetables as well as diverse plant life needed for insect pollination. The other ag trend in urban hubs is the reintroduction of backyard chicken coops and terrace beehives. The underlying ecomodern reason for increasing animals and other species, and reintroducing agriculture back into the urban landscape, could be the increase in sustainability factors: reducing environmental impacts, encouraging greater biodiversity, and maintaining a healthier microbial community. As urban populations soar upward of 730 million people—about 8.7 percent of the global population is projected to live in cities with at least 10 million residents by 2030[32]—how we integrate, inter-relate, and cohabitate more closely with animals will impact the biodiversity of our collective urban biome.

One thing urban planning and architecture has to keep in mind is that we humans and our pets are the species that help populate the inert surfaces of our built environments. And to keep our buildings and each other healthy we are to embrace rich, diverse ecologies into our construction that foster symbiotic biological systems to thrive.

Some of the ways we can start to rewild our cities, our buildings, and our neighborhoods is by integrating ecosystems in thoughtful, and perhaps even surprising ways that will attract new species of pollinating insects, or provide microfarming built right into an apartment complex. Architecture and urban design research group Terreform ONE is one group who work toward enhancing the urban environment with green technologies. One of the latest proposals of the nonprofit group is to create an eight-story-high monarch butterfly sanctuary, or "Lepidoptera terrarium," that would serve as the building's facade and line its atrium. The suggested location? The Nolita neighborhood of New York City. According to Terreform ONE's website, the proposed integrated biodiverse facade is "not just a building envelope, the edifice is a new biome of coexistence for people, plants, and butterflies."[33]

Hot Pets in the City, Runnin' Wild and Lookin' Pretty

Take this challenge: If you live in a major urban city, go for a walk and look around you. How many animals do you see? Birds and pigeons, check. Dogs on leashes, check. Squirrels? Possibly. How about a stray cat, or even a rat? If your list is pretty short, you're not alone: a lack of diversity in animal populations has been the trend for decades, where most cities have very homogenous animal life—most of which is probably on a leash.

Back in the eighteenth and nineteenth centuries, pets and domesticated animals were once fellow residents, roaming the city streets of New York, London, Paris, Berlin, and other cities with horse-drawn carriages, pigs, and chickens and other fowl that were kept in small city plots. But industrialization pushed buildings toward the skies, motorcars put horses back into their stables (the lucky ones as least), and the pigs and chickens slowly transitioned to being raised in commercial farms or factories. Modernity pulled the plug on animal biodiversity in the city.

Pigeons and squirrels aside, the animal population that has been on the rise, at least in the United States, are pets. Sixty-seven percent of US households, or about 85 million families, own a pet according to the 2019-2020 National Pet Owners Survey conducted by the American Pet Products Association (APPA).[34] We don't have a census of how many of these pets live in urban areas, but it has been estimated that of the roughly 146 million cats in the United States, about 50 percent of them are feral or unowned.[35] Philadelphia, the city of brotherly love, claimed in 2019 that it had nearly half a million stray or feral cats roaming its streets.[36] That's one example of wildlife in the city that could be spreading its microbes freely.

Like we learned in chapter 2, our companion species have their own microbiome, and we have learned that when they cohabit with us indoors, we are sharing our microbes. And we discussed that the one variable in every city is the migration of humans; walking or driving or traveling to and fro, everywhere we go, our microbes hitch a ride, spreading a microbial cloud. The same is true of our pets: when they take their walk or go for a run in the park, their clouds come along for the ride. Unfortunately, when compared to the number of other species in town (microorganisms, insects, and even plants and trees), the number of city pets still may not be enough to help maintain a healthy biodiversity in our asphalt-dominant lands.

Micro-Straphangers

We know the importance of underground microbes and their role in keeping our soils rich, and potentially helping with climate change. But what about the trillions of microbes in the Underground—those dwelling in our metro subways? While we are still exploring and documenting the various colonies of microbes all around us, several studies have directed their attention to a transient yet controlled environment—that of the urban subway. Subways are the pulse of any city, moving millions of

Pets in the United States

Number of US Households That Own:
A dog, 63.4M
A cat, 42.7M
Freshwater fish, 11.5M
—American Pet Products Association's 2019-2020 National Pet Owners Survey[37]

Pets Make You Healthy

Some of the benefits to living with pets include decreased stress, decreased blood pressure and levels of cholesterol and triglycerides, as well as decreased feelings of loneliness. Pet ownership can increase your longevity and desire for physical activity and socialization.
—Centers for Disease Control and Prevention[38]

riders and their microbes every day. Tokyo's Metro, the world's busiest subway system, averages 6.84 million people a day.[39] Now add the trillions of microbes also catching a ride with these people, and you could imagine the subway car as just one big microbial breeding ground. *Eek*, right!?

Years before COVID-19 hit, a few studies by ambitious researchers looked at what makes up the community of our mass-transit microbes, and not all microorganisms were bad. A study in 2018 of the microbial community of the Hong Kong subway system revealed that while each line had its own common species, it was mixed and dispersed throughout the day by its commuters. The most common bacteria found? Skin bacteria. Makes sense, since our skin is our largest organ and the exterior of our microbiome home.

Conducted by the University of Hong Kong, the research involved volunteers who were asked to first clean their hands, then enter the subway car and hold the handrails while riding for at least thirty minutes. The samples were swabbed from the palms of the volunteers, and were taken over the course of a

day from volunteers who rode in the morning or the evening. What was interesting is that during the morning rush hour, the researchers found about 140 species of microorganisms, but by evening many had dispersed, and a mere 48 species covered the entire line. What happened? The millions of humans jumping on and off the train all day. Each route's particular microbial community was present at the start of the day but shrank as the day went on, probably hitching a ride with a straphanger to their final destination. But you might be happy to know that even though the researchers concluded that the subway surfaces can influence our skin microbiota, most of the bacteria found were harmless to humans.[40]

New York's famous subway station system microbial mapping took another route—researchers set out to map all the known bacteria hanging out in the subway stations as well as the trains. The 2015 study was the first to map the spread of bacteria in a mass transit system. Led by Weill Cornell Medical College in New York, the research collected DNA samples from all the transit system's exposed surfaces—from the turn-

The World's Busiest Subways Transporting Us and Our Microbes (Based on Annual Passenger Rides)

Tokyo, Japan, Metro—
3.16 billion rides

Moscow, Russia, Metro—
2.4 billion rides

Seoul, South Korea, Metro—
2.04 billion rides

Shanghai, China, Metro—
2 billion rides

Beijing, China, Metro—
1.84 billion rides

New York City, United States, Subway—1.6 billion rides

Paris, France, Metro—1.5 billion rides

Mexico City, Mexico, Metro—
1.4 billion rides

Hong Kong, China, Metro—
1.32 billion rides

Guangzhou, China, Metro—
1.18 billion rides

—ThoughtCo.[41]

stiles, wooden and metal benches on the platforms, stairway handrails, trash cans, and kiosks to the inside of trains, including seats, doors, poles, and handrails, in twenty-four subway lines in five boroughs. They named their transit system microbial map the "PathoMap" (short for "pathogen map").

Like the Hong Kong study, the New York team found that most bacteria were harmless, but concluded that the heavily traveled subway could be a proxy for the city's population and to large-scale health management for New York.[42]

The Atmosphere Is Alive and Teeming with Microbes

While we have been illuminating the invisible neighborhoods of microbes on built surfaces on the streets and underground, we need to be mindful that for billions of years, microbes have been residents of every inch of the planet—including the atmosphere. Yes, what's below is above. We are now learning that there is a microbiome high in the sky.

Two groundbreaking studies quantified the multitudes of bacteria and viruses are blown upward into the part of the atmosphere that lies between the area of Earth's weather systems and just below the stratosphere where airplanes fly. These flying microbial particles break loose from soil dust and sea spray, freely traveling skyward into the atmosphere where they can linger for days before being dropped back on to the Earth's terrain.

A 2018 joint study by researchers at the University of British Columbia, Vancouver, the University of Granada, Spain, and San Diego State University, California, estimated that more than 800 million viruses are deposited per square meter into the atmosphere every day—that's about twenty-five viruses for each person in Canada.

In the samples collected, between 60 percent and 100 percent of the bacteria were alive, and they typically lived in the sky for five days or more. The lower atmosphere seems to be the most populated, and it is bacteria that win the microbial version of

Survivor, as they have the coping mechanisms to deal with UV radiation, lack of water, and high levels of ozone. They too feed on unusual things to survive–their sky buffet is the carbon compounds in the atmosphere.[43]

Hitching a Ride on the Highway in the Sky

Researchers have found that not all floating microbes linger in one area. Some viruses have been found to catch a ride with the planetary wind systems, jetting them across the globe.

Microbes have been traced over vast distances, and researchers have been able to use computer modeling of the Earth's atmosphere to see how microbes may be blown between continents. In a 2011 study, scientists from the United Kingdom and Switzerland simulated a flight path for intercontinental microbial travel, questioning the potential of these airborne microbes to colonize in new lands and perhaps even spread diseases.[44]

The 2018 joint study that identified the microbiome in the sky used viral detectors placed in areas of the Sierra Nevada mountains in Spain, and they found that billions of viruses and tens of millions of bacteria per square foot were falling from the sky and blanketing the mountain sites. Viruses were the dominant skydivers, depositing at greater rates than bacteria did. The researchers concluded that rain and Saharan dust storms were responsible for sending these airborne microbes back to Earth.

Another part of the study delved deeper into the microbial mix that managed to become airborne from sea spray. Using a thirteen-thousand-liter tank of water from the Pacific Ocean in a filtered indoor atmosphere, the researchers were able to simulate the natural conditions of wind and waves and sampled the resulting sea spray. When they did a genetic test of the sea-cloud sample, they found that *Corynebacterium*, a type of skin bacteria that was also found on the skin of people living around the university in San Diego, were more prominent in the sample from the spray than the samples taken from the water directly. Thinking what I'm thinking? Could it be that these sea-sprayed floating microbes make their way to land, and human contact? Or could it be the reverse: our human microbial clouds travel to other terrains besides our immediate vicinity, like to the seaside? The work of the study was intended to look at the possible connection of airborne clouds of microbes to the Earth's climate. More research is needed, but if microbes released from the ocean could be used to help manage global temperature, we may one day see some new "clouds" of carbon-eating flying microbes working their climate-action magic.[45]

As the research continues, we will learn of ways to manipulate microbial communities in the sky to rewild our urban ecosystems, as well as a host of other things we humans need help with—such as fighting climate change, recycling waste into new materials, or fermenting new foods.

Power to the Urban Holobiont

Before we jump planets, we need to note that there are several urban initiatives and proposals that plan to put microbes to work in our 'hood. Next generation restoration projects look to incorporate values of "microbiome-inspired green infrastructure" (MIGI) that influence the microbial diversity and exposure of urban residents. A 2018 proposal submitted to the Healthy Urban Microbiome Initiative (HUMI) entitled "Walking Ecosystems in Microbiome-Inspired Green Infrastructure" recommended the benefits of both greening urban centers and encouraging microbial diversity to improve urban community health. The team suggested that food for-

Disease in the Jet Stream

One of the interesting possibilities of the 2011 simulation model was that only the smallest microbes travel freely between continents within a span of a year, increasing the microbial diversity where they land and colonize. While most studies on atmospheric microbes believe them to be mostly harmless, past research has shown that outbreaks of meningitis in areas like the Sahel region of Africa, and conditions like foot and mouth disease, have been linked to airborne microbes.[46]

Naturally, this finding begs us to question if this was the case with COVID-19. While we are writing this book during the COVID-19 pandemic, there is still some speculation about how the virus has rapidly spread, and if there was more risk beyond physical contact. The World Health Organization in March of 2020 warned that the coronavirus may also be airborne.[47]

Some scientists have even speculated that the COVID-19 virus may have been carried between countries at similar latitudes by a jet stream through the atmosphere.[48]

aging in urban community gardens and creating green walls with diverse flora, fauna, and microbiota may also help with air and noise pollution.[49]

Now, About That Kitchen Sink . . .

One of the more ecomodern fixtures in urban cities of late has been the compost bin, or in some communities, the shared compost station. Compost bins are used to store most organic "trash"—from used coffee grinds to eggshells to banana peels, and every bit of fruit and veggie you can imagine—and they are phenomenal microbial waste recyclers. Microorganisms such as bacteria, fungi, and actinomycetes act like your trash recycler, breaking down most of the organic material inside the compost bin. More eco-minded people living in the city or the suburbs use compost bins for reducing waste and generating good nutrients for enriching the soil of their gardens or house plants.

New designs are moving beyond the trash-can look of most compost bins into a seamless integration into the home kitchen, including the sink. Back in 2011, Philips Design, headquartered in the Netherlands, released their vision of the kitchen of the future as part of their "Microbial Home" system. The concept proposed a bio-digester kitchen island, in which bacteria digested organic waste such as vegetable peelings and turned it into methane gas for cooking and lighting. The dehydrated sludge from the digester would be used as compost.[50]

For now, though, there are versions of the home compost bin built right into the drainpipes of your kitchen sink. Similar to the sink garbage disposals that grind food waste into small particles that are flushed out of the disposal chamber when you run water into your sink, there are now sink-integrated composting systems that attach to your existing plumbing and turn your food waste into compost . . . with the help of friendly microbes, of course.

Whether under our kitchen sink or hanging in our subways, the interdependency of microbiomes among humans, animals, plants, and microbes is a relatively new area of research, especially in an urban setting. What we have yet to realize is the aftermath of one invisible viral force called COVID-19 and its unimaginable power to shutter most dense urban centers during quarantining. We have yet to know the full impact of the pandemic on our urban microbiomes, but we at Alice are hopeful that the findings will steer us toward new ideas and practices for rewilding our urban microbial cities to work harder to sustain our health. That's where the biotopians come in—the next generation of architects, designers, urban planners, farmers, food, and material manufacturers that are using biology, and microbes as their tools. Time to meet them!

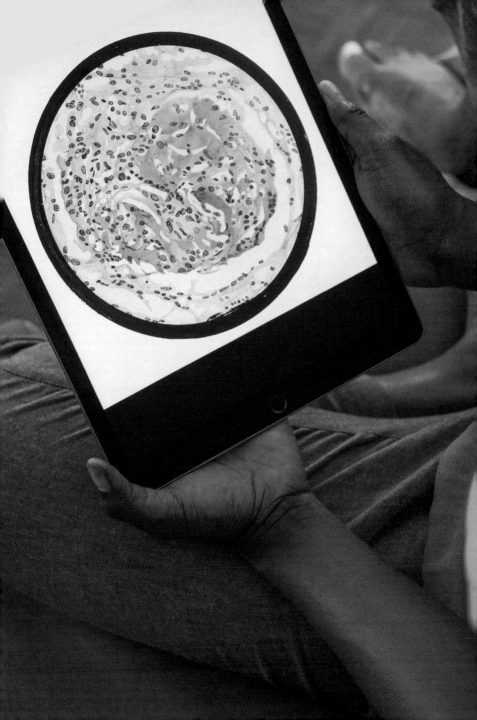

Here Come
the Biotopians

Biodesign with Microbes

Bioimagine
the Future

We are on the cusp of a new frontier, and it is one that is biodesigned. As we better understand the molecular biology of life, there's really no limitations as to what we can get living things to do. You see, living things can be used as factories to create any molecule. Yes, any molecule. And since molecules are the basis for everything we construct, biodesigners are engineering technologies for every single purpose they can imagine—constructing new materials, new substances, new foods.

Not since the dot-com boom has there been such excitement about future pockets of rapid growth. Welcome to the "bioeconomy." The bioeconomy is a term experts use to describe all the new and innovative ways we are producing food, products, and energy, using biology as our tools, including the handiwork of microbes.

Ten years ago, in his speculative *New York Review of Books* essay "Our Biotech Future," theoretical physicist Freeman Dyson imagined a time when life-forms, not just startups, might be hatched in garages; when reptile breeders could fashion designer lizards; and children could create their own playthings. He posited a "bright future for the biotechnology industry when it follows the path of the computer industry, the path that von Neumann failed to foresee, becoming small and domesticated rather than big and centralized."[1] We at Alice in Futureland couldn't agree more. As biotechnology makes its way into our products and our homes, we will have at our fingertips the means to self-produce new medicines, materials, foods, and energy sources using the elegant and industrious biological processes of bacteria.

Biodesigning the Code of Life

We know bacteria are smart, social, and even altruistic, but who knew that an editing strategy they use would become our most powerful tool to precisely modify, and redesign, the genetic code of life.

CRISPR is the revolutionary gene editing tool bacteria use for immune defense, whereby they take "snippets" of invading viruses, place the clippings in storage, and pull them back out to use against other invading viruses. Another way to understand the process is to think of how you use software for creating documents or editing films: cut, copy, and paste. CRISPR is bacteria's editing software, an immunological defense mechanism employed against invading pathogens. We owe gratitude to our CRISPR critters for teaching us this gene-altering process, as its efficacy and ease of use is helping us accelerate innovation across diverse industries, including health care, agriculture, and clean energy.

CRISPR is as elegant as it is precise, enabling us to edit the genetic code to do what we need it to do. But we are just at the early stages of using this immense editing tool for our benefit. There are other microbial superpowers we're harnessing for developing new foods and new materials, such as fermentation, which turns various forms of waste into new resources for foods or fuels.

No doubt we're moving away from the crass industrial age of producing with chemicals into a biological future in which microbes are our greatest performers. We may one day use the prefix "bio" for just about every occupation that creates or manufactures goods, services, materials, and natural resources. But for now, we are all biotopians, imagining a healthier, more equitable, more plentiful world engineered by microbes.

CRISPR Health Care

Although CRISPR was discovered in 2012, the first human clinical trials began in 2019, using CRISPR/Cas9 to fight cancer and blood disorders in people. How it works: Researchers remove some of a person's cells, edit the DNA, and then inject the cells back in—armed with the tools to hopefully fight disease. In health care, CRISPR holds the promise of eventually curing most of the over six thousand known genetic diseases.[2]

Putting CRISPR to Work

CRISPR techniques have already been applied to modify yeast to make biofuel, and in the food industry, to engineer probiotic cultures and vaccinate industrial-level microbial cultures, like those used for making yogurt, to protect the cultures from viruses.

In agriculture, CRISPR is used to help improve the yield of crops, make them drought tolerant, and provide greater nutritional properties. Pairwise Plants is working with CRISPR to create new crops and modify existing ones to require fewer resources to grow.[3]

The real innovations here over the last twenty years have been three big things—one, our ability to sequence DNA to read information from living organisms. All of the code that makes up the instructions to tell us who we are and how to make another one of us is stored in the three billion letters, the As, Ts, Cs, and Gs in your DNA. And that's true of every organism on the planet. The second big thing that's happened is that we can write DNA, so we can chemically synthesize it. We mash ATC and G a lot on a keyboard, we send it off, it gets synthesized as a little film of powder that comes back in a FedEx box, and then we can put that into a microorganism and make a little bit of that code. In our case, a protein or a structural protein like spider silk. And then the last thing is the tools and the computation around how we understand living systems around us and can use the direct copy

Biodesign Is Alive

from nature, not just as inspiration but understand it directly and replicate it without the organism.

—Dan Widmaier, founder and CEO, Bolt Threads, interview with Sputnik Futures, 2019

Bio Is the New Platform for Creativity

"Biotope," in biology and ecology, refers to a region, habitat, or vital space associated with an ecological community—a living environment. Biotopians are a part of a growing movement of scientists, artists, and designers who want to rethink how we build, well, everything!

The exciting thing about the biotopian movement is that it partners scientists and designers. Artists and designers draw upon the expertise of biologists; biologists benefit from the imagination and outside thinking of artists and designers. By harnessing organic processes and materials, biotopians will rethink how we create buildings, our clothing, our beauty products, and even our food. The rise of biodesign puts designers in a thought-leadership position. The hope is a world where sustainability can be manifested by altering the DNA of living organisms. In the new world of microbial design, we will need biotopians to narrate and communicate the new story for the public-at-large. In fact, the Museum of Modern Art's architecture and design curator Paola Antonelli has been on the biodesign frontier since 2008 with the curated show *Design and the Elastic Mind*.[4]

Biodesign is the twenty-first-century equivalent of the UX (user experience) design. Computer programming was once performed only by tech heads. Few people even knew how to write code! But today, with basic block-based programming like Scratch and text-based coding being taught in grade school and the simplification of open-source software, the digital age is upon us. Similarly, the technologies related to bacteria manipulation—operations that until a few years ago were very expensive and resided only in large laboratories—are now increasingly accessible to students, designers, and citizen scientists. We expect a new wave of responsive solutions using microorganisms and bacteria.

More Aesthetics, More Ethics

If Freeman Dyson's suggestion that in the future biotech will be domesticated and that there will be do-it-yourself kits for gardeners to use gene transfer to breed new varieties of roses and orchids and children will play biotech games that have real eggs and seeds rather than images on a screen, then the role of bioethics will definitely take center stage. As biodesign scales up, ethical, environmental, and political questions will arise. Once genetic engineering gets into the hands of the general public, it will give us an explosion of biodiversity.

Encouraging a design future in biotechnology is the heart of the Biodesign Challenge (BDC), a student education competition that integrates synthetic biology and design with a purpose to promote the right dose of creativity and ethics. Students from classes in art, design, and biology are paired with subject matter experts, and tasked to think about the future of biotechnology. At the end of the semester, the top teams from each of the universities involved present either in New York City on stage at MoMA or Parsons School of Design or virtually to compete for the top prize. The BDC panel of judges is comprised of architects, scientists, biologists, philosophers, lawyers, and financial experts.[5]

In an interview with BDC executive director Daniel Grushkin in 2020, Sputnik Futures explored the intersection of biology, design, and ethics.

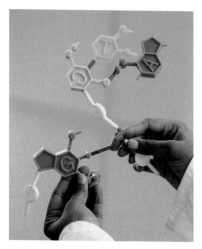

I also encourage folks to really think about art because art plays a really important role in this. Art asks a lot of the questions about why we do the things we do, and what meaning they have, and what our life means when we have this in it. All those questions are really important when we're starting to reimagine what's possible with the designed world.

SF: Now, as a designer, do I need to be a scientist to involve myself in this?

DG: You don't need to be a biologist to be a bio-designer. In fact, what you need to be able to do is cross those disciplines. You need to be someone who is able to speak the language of biology so you can speak with biologists, but you also need to speak the language of design so that you can design as a designer.

We're not just talking about better packaging for our products. We're talking about reimagining some of the basics of what it means to be human, some of the most foundational things about our lives, our rituals, the way that we think about death, the way that we think about life, the way that we think about our relationships with each other, the way that we identify ourselves. All of these are really under question when you start to think about how biology can affect those states of being, those relationships.

—Daniel Grushkin, executive director, Biodesign Challenge

Focus on Biodesign

Biology is becoming part of pop culture as we embrace new foods, materials, and home solutions that are inspired by or even use microbes as their building blocks. New media publications, books, and exhibitions showcase the interconnectivity between design and synthetic biology. For example, *Nature–Cooper Hewitt Design Triennial*, co-organized with Cube design museum, presented the work of sixty-two international design teams with projects ranging from experimental prototypes to consumer products, immersive installations, and architectural constructions. Running from May 2019 to January 2020, it featured collaboration between scientists, engineers, advocates for social and environmental justice, artists, and philosophers. A highly innovative exhibition, it explored engagements with nature, climate change, and ecological crises, imagined with science and technology.[6]

Biodesign is still a growing field, with several university departments already offering courses in it, such as Stanford Byers Center for Biodesign, California; John Hopkins University's Biodesign Program,

Biodesign Challenge Winners: Pseudofreeze

The Cool, Humanitarian Promise of Bacteria:

A group of students from the Universidad de los Andes in Bogota, Colombia, won the 2019 Biodesign Challenge with their design of a canister lined with self-cooling bacteria to help transport vital medicines that require refrigeration in areas with high temperatures and limited access to refrigeration. They looked at a bacterium called *Pseudomonas syringae* that has the special property of causing water to freeze using one of its proteins. The students lined the canister with a chamber that holds water; when it's time to transport the fragile vials of medicine, the bacteria's protein powder is added, freezing the water and keeping the vials cool.

Death on Mars

A group of students from University of Technology in Sydney, Australia, created a project called Death on Mars that imagined the future of the cemetery. Their idea was based on the research of a Harvard lab that demonstrated that you can store video images, encode them in DNA, and then insert them into bacteria, and later pull it out of that bacteria and reconstitute the video image. The students imagined embedding personal letters, photos, and other personal effects into DNA code, and then put that code in plants or trees, so when we go to a "cemetery" here on Earth, or eventually Mars, it will be a grove of trees with the DNA of people who passed away.[8]

Biodesigned Spaces

One issue of the online magazine *Biodesigned* featured leading thinkers who explored the built environment and how it can be reimagined, redefined, and reconstructed. One architect proposed the idea of making the construction industry totally circular by feeding demolished buildings to fungus, which then creates new materials.

—Biodesigned[9]

Baltimore, Maryland; and schools such as Arizona State University, Harvard Medical School, and University Arts London, to name a few.

Biodesign is also taking over the private sector, with companies like Ginkgo Bioworks, who have been at the forefront of design by biology for the past ten years. The company designs custom organisms for multiple industries, using what they call "foundries" to scale up the process of organism engineering. For consumer goods, they use the process of yeast fermentation to create cultured ingredients for new and novel flavors and fragrances, enabling the production of both renewable, great tasting, and great smelling products. Think of it like a microbrewery: ingredients that are brewed by engineered yeasts in large vats. The organism engineers at Ginkgo Bioworks use gene sequencing, drawing from plants and other organisms, and synthesizing and engineering them into the genomes of yeasts. As these engineered yeasts grow and ferment sugars, they produce the fragrance and flavor compounds. One of their recent developments is in high-quality cultured cannabinoids, created in partnership with Cronos Group.[10]

The Fabric of Biodesign

Biodesign is being incorporated into the practices of artists, architects, and fashion designers to tackle—and humanely solve—today's sustainability issues. One industry that is stepping up is fashion, and it should since the UN Conference on Trade and Development (UNCTAD) considers it to be the second most polluting industry in the world (behind oil). According to the UNCTAD, the fashion industry annually uses around 93 billion cubic meters of water—equivalent to the water usage of five million people. Your clothing, and every time you wash it, is adding to the half a million tons or more of microfiber (equivalent to 3 billion barrels of oil) in the ocean every year. Since clothing and textiles are made in all parts of the world, the fashion industry is responsible for more carbon emissions than all international flights and maritime shipping combined.[11]

Designers, textile manufacturers, and retailers from denim to couture are taking notice

Color Made in Microbes

PILI uses enzymes and microorganisms to transform sugar into vibrant, renewable pigments for textiles, rather than the traditional petrochemical dyes that pollute our water streams. One kilogram of PILI dye saves one hundred kilograms of crude oil and ten kilograms of toxic chemicals, while using five times less water.[12]

Faber Futures, the London-based studio led by designer Natsai Audrey Chieza, uses the soil-dwelling bacterium *Streptomyces coelicolor* that produces a pigmented compound to dye fabric and garments in patterned hues of pink, purple, and blue. The range of bio-dyes were designed in a reversible silk coat commissioned and acquired by the Cooper Hewitt in 2019.[13]

Algalife develops natural and healthy pigments and fibers from the same microorganisms—algae. Algae are a valuable substance of proteins and vitamins, is anti-inflammatory and antioxidant, making the dyes and fibers skin-healthy too.[14]

and have been working on solutions for "greening" fashion. Constructed using synthetic biology, microbes are naturally becoming our new smart materials, reshaping our notion of fashion, food, and industrial design. Synthetic biology is reinventing advanced materials by engineering bacteria, leading to new ecomodern design practices. Due to the backlash on the waste produced by fast fashion, biodesigners are harnessing microbial systems to create eco-friendly fabrics, dyes, and materials, redesigning the future of our clothes and challenging throwaway culture. Beautiful, sustainable dyes made from bacteria are coming from companies such as PILI. Natsai Audrey Chieza, a London-based designer and founder of Faber Futures, has harnessed bacteria to dye fabrics without water.

Biodesigned textiles are also coming soon to a store near you. For example, designer Stella McCartney, a forerunner in tackling sustainable fashion, teamed up with the Biodesign Challenge in 2018 and animal rights charity PETA to invite students to develop proofs of concept for a sustainable new "vegan wool." The production of sheep's wool is considered more polluting than that of acrylic, polyester, spandex, and rayon fibers. According to the "Pulse of the Fashion Industry" report, wool production has a poor pollution ranking based on cradle-to-gate environmental impact per kilogram of material.[15] Recognizing this, "WOOCOA," the winning team from La Universidad de Los Andes, combined coconut fibers left over from the food industry with hemp fibers and applied naturally occurring lactase enzymes found in oyster mushrooms—which are bacterial lactase. These were used to break down the fiber's roughness, resulting in a "new bio-fabricated wool, which is flexible, elastic, easy to dye, and wicks away humidity."[16]

Better still, some companies, such as Bolt Threads, are designing fully biodegradable fabrics. Founded in 2009, their aim is to develop better materials for a better world. You may have heard about their pioneering fibers derived from spider's silk. But what Bolt Threads did in 2012 was recreate the high tensile strength, elasticity, durability, and softness of spider's silk using yeast. Their process starts with bioengineer-ing the silk's proteins, putting the new genes into yeast, and then producing the engineered protein through fermentation using yeast, water, and sugar. The final step is isolating and purifying the new engineered silk protein, which is then spun into fibers, likened to rayon and acrylic but silk-derived. Bolt Threads knits these fibers into fabrics called Microsilk. The entire process has a significantly less environmental impact than traditional textile manufacturing—and a garment made with Microsilk can potentially biodegrade at the end of its life.

The team at Bolt Threads has even found a second life for the protein waste that comes from the fermentation process. This isolated unhydrolyzed silk protein that they called B-Silk Protein is also produced via fermentation and is a key ingredient in Bolt Threads' Eighteen B skin-care brand.[17]

The vision for a global wardrobe made and dyed by "living factories"—such as mycelium, bacteria, yeast, and algae—is that of the biofabrication pioneer Suzanne Lee, who coined the term "biocouture" in 2004 to describe her leading research in growing

Microbial Enzymes

These designer enzymes (catalysts for biochemical processes) can be produced through fermentation techniques that are cost-effective and require less space and time to make. The numerous applications for microbial enzymes include:

· Food: production of glucose syrups, crystalline glucose, high fructose corn syrups, maltose syrups, etc.

· Detergent: used as additives to remove starch-based stains.

· Paper: used to reduce starch viscosity for coating of paper.

· Textiles: used for warp sizing of textile fibers.[18]

Microbial Leathers

Artists and designers are exploring the potential of microbial leather as a sustainable, ethical, and environmentally positive alternative to animal-sourced leather. Microbial leather is made from cellulose nanofibrils spun by bacteria and yeast.[19]

Materiability Research Group has an online tutorial for growing this leather alternative, using ingredients like one kombucha culture, 200 ml of apple cider vinegar, 240 g of granulated sugar, three black or green tea bags, and water. The grown material can be treated and dried into a sheet form or it can be molded around a form during the drying process.[20]

clothing using living organisms. In 2006 she "grew" a denim jacket from microbial cellulose, expanding into various biofabricated materials. A year later she published the groundbreaking book *Fashioning the Future: Tomorrow's Wardrobe*, which mapped out the new wearable technologies for fashion and textiles. She continues her bio-microbial innovation path today with Biofabricate, a platform she founded for collaboration in design and biology to grow the future of sustainable materials for consumer products, working with designers, brands, startups, and investors.

As we enter a new material age, biodesigners will be driven by creativity, innovation, and environmental responsibility. By utilizing materials that use living systems, they are manufacturing future consumer products by harnessing microorganisms. We are challenging traditional processes that require the use of animals (their fur or hide) for the sake of putting more clothing or accessories into the world, because with biotechnology, we will no longer have to rely on this outdated model. Synthesizing nature's materials with new functional properties will lead us into a materialistic future with an ethical and regenerative vision.

What's on The Menu? Hydrogenotrophs

Are you a vegan, plant-based eco-modernist? Due to climate change there is an ongoing virtuous debate about plant- versus meat-based diets. Climate activists want to save the environment one plate at a time. Consider this: What if a third party shows up at the table?

You guessed right: microbes. What if our well-meaning foodie friends stop debating plant or animal and instead start embracing abundant, sustainable, unicellular ingredients? For one thing, it would make for an interesting Instagram feed, and, more important, it would further the food security and environmental conversation.

So, what are we talking about when we say unicellular food? Think craft beer, brewing microbes through fermentation. Multiplying microorganisms to create the special of the day. Fermentation can scale up and could feed the world.

MYLO by Bolt Threads is a biofabricated leather alternative, engineered by mycelium, the vegetative part of a fungus or fungus-like bacterial colony, consisting of a mass of branching, threadlike hyphae that grows in soil, creating a web that breaks down organic matter and provides nutrients to plants and trees. MYLO will become available to the world through the support of consortium partners Adidas, Kering, Lululemon, Athletica, and Stella McCartney.[21]

TÔMTEX, a Biodesign Challenge 2020 finalist, is a leather "alternative" made from food waste like seafood shells mixed with coffee grounds and microbes. The leatherlike material can be embossed with a variety of patterns.[22]

ALGIKNIT is creating degradable yarns from kelp. Kelp is a type of algae, and one of the most renewable organisms on the planet, absorbing CO_2 and working against global warming and ocean acidification.[23]

Grow Your Own . . .

MASK: A prototype by designers Garrett Benisch and Elizabeth Bridges used bacteria to grow what they call the "Xylinum mask" out of bacterial cellulose, a by-product of a common bacteria called *Xylinum acetobacter*. The designers offer some home DIY tips to grow your renewable and biodegradable mask using a few simple kitchen ingredients: water, tea, sugar, and a small bacterial sample from unflavored kombucha.

—*Fast Company*[24]

For example, Culture Biosciences is a company that is described as a virtual fermentation lab. The lab has the ability to harness precise gene-editing technologies like CRISPR combined with algorithms to design a microbe to make new food products. Companies are utilizing labs such as Culture Bioscience to brew up vegan versions of everything from meat and cheese to eggs. Now, *that* is a twenty-first-century omelet!

As we discussed, some bacteria turn CO_2 into valuable fuels and some bacteria can transform CO_2 into proteins suitable for human consumption. Gourmet bacteria are called hydrogentrophs. Hydrogentrophs could produce a given amount of protein on one-twentieth of the land that a traditional farm requires, and research suggests there's an enormous potential for creation of planet-friendly protein using bacteria and other microbes.[25]

Solar Foods is one company delivering a climate-friendly, protein-rich food made from water-laced bacteria, electricity, and air. Its Solein is protein-rich powder developed in collaboration with VTT Technical Research Centre of Finland and the Lappeenranta University of Technology. There are no animal proteins here—Solein is made by applying electricity to water to release bubbles of carbon dioxide and hydrogen captured from the air. Living microbes are added to the liquid, making the protein that is dried to what may be the most economical and sustainable protein-rich powder. There are great plans for feeding the world with Solein powder—up to two billion meals a year! It would be wonderful news for the planet as well as our impending food shortage: it takes 1,550 times less water to produce Solein than the equivalent amount of protein in beef, and Solein is ten times more efficient than soy in terms of usable protein yield per acre. And here's another number to consider: microbes can make a protein powder that is "100 times more climate-friendly" than other food.[26]

Munching Microbes

Humans have been munching on microbes since before Neanderthal times—actually, from the first moment we ever ate. The earliest members of the genus Homo evolved around 2.8 million years ago, but there have been contradictory views over what our earliest ancestors ate. Some camps claim they were purely meat eaters, but plant remains found in teeth and, more recently, feces of Neanderthals collected by a team of archaeologists at the El Salt excavation site in Spain show that Neanderthals did indeed eat their veggies.[27] And guess who was living on those vegetables? Microbes!

Yes, our ancestors bioengineered their foods with yeast. Humans have been using the fermentation process for thousands of years, letting yeast do the work to make things such as fermented alcoholic beverages from fruit, honey, and rice as early as 7000 BCE. Around 10,000 BCE humans took advantage of the tropical heat and its fermentation effects on dairy products, producing the first cultured milk in goat bags draped over the backs of camels in North Africa.

However, it wasn't until Louis Pasteur connected yeast to the process of fermentation in 1856 that we expanded our understanding of what was happening in the process, opening the door to later discoveries of its health benefits. It was around 1910 when a Russian bacteriologist named Elie Metchnikoff noticed that Bulgarians lived roughly eighty-seven years on average, which was extraordinary for the early 1900s, and he attributed this to the fermented milks they drank. He named the bacteria found in the Bulgarian fermented milks *Bulgarian bacillus*, which eventually was renamed *Lactobacillus bulgaricus*, the main bacteria in your cup of yogurt. Large bodies of research for the last five decades have been examining the health benefits of friendly

bacteria in our foods, which influences our gut microbiota, and all the systemic health benefits that we discussed in chapter 3.[28]

Fermentation is perhaps one of the oldest methods of food preservation, and almost every culture has its favorite fermented dish or beverage: kimchi in Korea; natto and miso in Japan; tempeh in Indonesia; kombucha in China, Vietnam, Korea, Japan, Russia, and Ukraine; kefir from across Eastern Europe; sauerkraut from Europe, but an American favorite; and every loaf of sourdough is, you guessed it, made through fermentation.

Today, fermentation is our new food tech, creating alternatives to meat, fish, chicken, cheese, and more. Many modern companies are utilizing fermentation to make cow-sourced products such as chopped meat and milk without the cows. This new cultured, cellular-based food tech boom is what venture capitalists, economists, environmentalists, and more think will help feed our exploding global population— upward of 9 billion people by 2050. You just need to follow the money (the investment money, that is) to get a sense of how big this new food production will be: according to UBS, the market for plant-based proteins replacing meat alone could expand from just under $5 billion today to around $85 billion over the next decade.[29]

The key to the bioengineered process of making Beyond Meat and other plant-based foods is the microbial enzymes. Biotechnology has created novel enzymes with a wide range of applications, including new food sources. Microorganisms such as bacteria, yeast, and fungi, and their enzymes are widely used for improving the taste and texture of foods, and they offer enormous economic benefit to industries. In fact, microbial enzymes are now the preferred source of enzymes compared to plants or animals since they are more cost-effective, offer consistency and stability, and use fewer environmental sources in production.[30]

New food production is a big part of the world's growing bioeconomy. Look closer at Beyond Meat, for example. It took ten years of development, yet it had one of the most successful IPOs in US history. When Beyond Meat went public on May 2, 2019, they had a $1.5 billion valuation. Less than three months later, it was worth more than $13 billion.[31]

The bioeconomy is considered today one of Europe's "largest and most important" sectors, estimated to have an annual turnover of around €2 trillion, with roughly 76 percent of the 18 million people working in agriculture and the food and beverage industries.[32]

Tech investors have been eyeing the food industry for the past few years, and accelerators in biotech that invest in startups have also been churning out new cultured foods and food ingredients that are made with the help of microbes. IndieBio is one accelerator that has impressive graduates from its world-leading program. Born in 2014 in San Francisco, California, its twice-a-year program gives fifteen early-stage biology companies seed money, helping the founders who enter their four-month intensive program turn their concepts into real products that they can pitch to investors. Adding new offices in New York City and London, IndieBio is one of just a handful of accelerators that are investing in the future of plant-based food—as well as bioengineered solutions for medicine, health, and beauty—catapulting these sustainable and healthy products into the mainstream.[33]

While this is still an emerging industry worldwide, you've undoubtedly noticed the skyrocketing popularity of plant-based foods in the last few years. The promise of reducing greenhouse gas emissions and eliminating the need to consume animals are the biggest drivers of this boom, and it's becoming evident in our grocery shopping carts. According to data from the Good Food Institute and the Plant Based Foods Association, in 2018 US retail sales of plant-based foods significantly outpaced overall grocery sales, and the plant-based meat category alone grew by 10 percent to be worth more than $800 million, with consumption of refrigerated grocery-shelf plant-based meat growing 37 percent as compared to conventional (real) cow meat, which just increased by 2 percent in 2018.[34]

The Impossible Burger by Impossible Foods—another success in the plant-based meat category—uses the DNA from the roots of soy plants and inserts it into genetically engineered yeast that, once fermented, recreates the flavor and color of a real beef burger. It replicates the heme, or leghemoglobin, found in animal flesh that is the catalyst for the chemical reactions that occur while a burger is cooking. Kosher- and halal-certified, the Impossible Burger "bleeds" like a real juicy piece of beef, and packs about nineteen grams of protein per average serving. How popular is it? As of October 2019, the Impossible Burger made its way onto the menus of over seventeen thousand restaurants in the United States—including Burger King's, as the "Impossible Whopper"—and other locations such as Hong Kong, Singapore, and Macao, and numerous grocery outlets such as Kroger's, Ralphs, Wegmans, and Fairway Market in the United States.[35]

I Scream, You Scream, We All Scream for Flora Protein Ice Cream!

While plant-based milk alternatives made with soy, rice, coconut, oat, hemp, and more have been crowding the refrigerated milk shelves, Perfect Day Foods delivers the real taste of milk—made with microflora. They take the DNA of cow protein genes to recreate an animal-free milk protein, giving the microflora they use the genetic "blueprint" of whey and casein proteins, which encourages the microflora to "graze" in large tanks on these plant-based cellular inputs and produce identical milk proteins to that of a cow. The result? A pure and better-for-you-and-the-planet whey protein that makes creamy milk, cheese, and ice cream without the "moo."[36]

In order to fast-track their mild-protein alternative technology into packaged foods, the founders of Perfect Day Foods teamed up with a product developer in the dairy industry and created a new sustainably focused consumer food company called the Urgent Company. Its first product: Brave Robot ice cream, which tastes like real dairy cream but is made from flora protein.

In the next decade we will see more food and beverage products come to market that offer alternatives to animal-based foods, eventually disrupting the food supply chain with more shelf-

IndieBio has graduated a few disruptors to date, reimagining more sustainable foods and beverages. It was at IndieBio that the first lab-grown meatball was served. Here are some other delights that came out of the biotech accelerator:

· Egg Whites: Clara Foods' Clara Whites replaces egg whites that can be used in baking recipes and even for making a cholesterol-friendly omelet.

· Cheese: Let's not forget our favorite dairy product—one already made by fermentation! Now the startup New Culture is taking it one step further using microbes to express dairy proteins and culture real dairy, animal-free cheese.

· Seafood: Finless Foods creates sustainable seafood using scientific cellular agriculture technologies, while New Wave Foods, female-founded in 2015, began perfecting the world's only algae- and plant-based shrimp.

· Meat and Poultry: Memphis Meats is creating beef, chicken, and duck from cultured cells in about four to six weeks. New Age Meats is developing cell-based healthy pork sausage, while Prime Roots uses fungi "superprotein" as the base of their high-protein products, including Koji Bacon.

stable, environmentally efficient, abundant, and, as the technology scales up, cost-effective options to help feed our rising global population.

We at Alice are excited with these new possibilities—and may we request a plant-based gelato as well?

The Plant-Based Pet Food Revolution

Humans aren't the only ones who deserve animal-free, protein-rich food options—our pets do too! That has been the vision of long-time food and biotech advocate, venture capitalist, and vegan rockstar scientist Ryan Bethencourt, who in 2017 founded Wild Earth, the world's first plant-based pet food company. A clean protein dog food was its first product, made of a yeast that contains a complete source of all ten amino acids your dog needs. Wild Earth calls yeast "tiny miracles" that are miniature protein factories—pumping out ultra-high-quality protein, growing easily at scale, and requiring radically few resources to produce.

According to Bethencourt, "The most dangerous ingredient in meat-based dog food is the meat itself. There have been over 180 recalls in pet food since 2009 and almost all of them are from the meat ingredients."[37] By serving your pet vegan kibble, you could reduce your and your pet's carbon footprint, which is significant—feeding pets creates the equivalent of 64 million tons of carbon dioxide a year.[38] Wild Earth's low-cost sustainable protein option is what caught the investment eye of Mark Cuban when Wild Earth pitched on ABC's *Shark Tank*, raising $550,000 from Mark. The company has since raised more than $12 million, with plans of expansion and education in the human, planetary, and pet health benefits of yeast-based vegan protein.

- Sweeteners: Joywell Foods is a food tech company that makes exotic fruit proteins through both plants and fermentation to provide sweetness without sugar.

- Infant Formula: Sugarlogix is developing yeast-based technologies to produce the rare, functional sugars found only naturally in human breast milk.
                        —IndieBio[39]

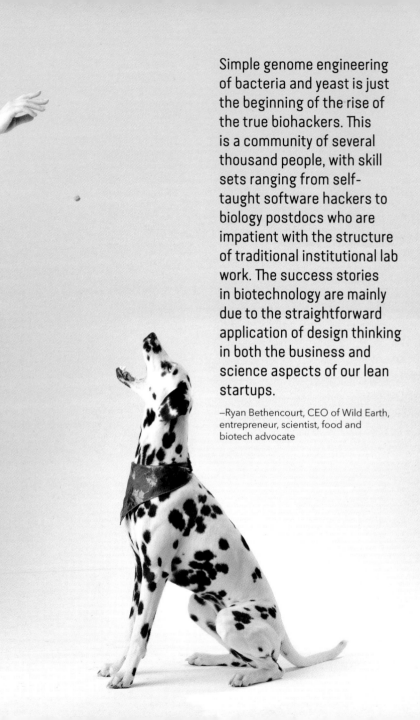

Simple genome engineering of bacteria and yeast is just the beginning of the rise of the true biohackers. This is a community of several thousand people, with skill sets ranging from self-taught software hackers to biology postdocs who are impatient with the structure of traditional institutional lab work. The success stories in biotechnology are mainly due to the straightforward application of design thinking in both the business and science aspects of our lean startups.

—Ryan Bethencourt, CEO of Wild Earth, entrepreneur, scientist, food and biotech advocate

A Is for Algae

Move over apple, algae are coming on to the scene, and in the future may be the first letter of highly sustainable, nutritious, and abundant food sources. Algae are broadly classified as Rhodophyta (red algae), Phaeophyta (brown algae), and Chlorophyta (green algae), and are categorized by size as either multicellular macroalgae (like seaweed) or single-celled microalgae that are eukaryotic, like green algae.

Microalgae grow in saltwater or freshwater environments, busily converting sunlight, water, and carbon dioxide to algal biomass. There is considerable interest in microalgae due to their industrious work making renewable energy (more on that in chapter 5), biopharmaceuticals, and, more recently, food ingredients.[40] Algae have been a food source since prehistoric times and are still part of the daily diet in Asia, where people consume red and brown algae.

Rich in numerous health-beneficial compounds such as omega-3 long-chain polyunsaturated fatty acids, microalgae are considered one of the most favorable sustainable sources of food in-

NovoNutrients is turning waste CO_2 into protein for aquaculture and animal feed and for plant-based or cultured "meats" for people. Also from the IndieBio network, NovoNutrients uses a proprietary microbial technology that is a platform for making nutritious, customizable inputs for the food system, supplying protein companies who make plant-based foods and feeds for animals. Its proteins are complete proteins, with all the amino acids found in meat, but less impact on agriculture, wildlife, and water.[41]

gredients. Microalgae, which can be bred in photo-bioreactors or fermenters, contain a high proportion of proteins (depending on the species), polyunsaturated fatty acids, beta-carotene, and numerous vitamins. There are over three hundred thousand species of microalgae, of which only fifteen groups are currently tapped for use in animal and human nutrition—with the potential to harness thousands more species.[42]

Another reason microalgae are an attractive food ingredient is its high omega-3 content. Essential for optimal brain function, and key to preventing cardiovascular disease, until now humans have often relied on fish for our omega-3 levels. Of course, the fish get their omega-3 power from the microalgae they consume, so why not go directly to the source? That's been the vision of food ingredient scientists who are developing the most economic and sustainable ways for introducing microalgae into our diet, but it's not that easy to overcome challenges such as oxidation.

Generally speaking, microalgae can be processed into either whole biomass or ex-tracted oil before being added to processed vegetable-based products for consumption. Large-scale biomass production has been ramping up with the use of industrial-scale fermenters. Key to the success of microalgae becoming part of our daily diet is putting microalgae ingredients in food products people will want to eat. Pasta may be one food that can garner some love for microalgae. Alver, a Swiss startup, has partnered with the Bühler food technology company to develop a microalgae-based pasta. In their test production, the team used golden chlorella microalgae with a 63 percent protein content to produce long and short pasta samples.

According to the European Algae Biomass Association, there are more than two thousand companies worldwide that are active in the production or processing of microalgae, paving the way for greater opportunities to eat your algae . . . or maybe even drink it.[43] At least that's the hope of three women entrepreneurs of FUL Foods who launched a microalgae-based beverage as the flagship product in a line of microalgae-centered products.

The team created a protein-rich "miracle food" ingredient they call FUL that also provides immune-boosting, natural nutrition derived from sustainable "plant positive" microalgae with a refreshing taste.

Spirulina Latte, Anyone?

Other popular microalgae found in beverages, supplements, and powders are chlorella and spirulina. Both pack great nutritional benefits, such as managing blood sugar and the risks that could cause heart disease.[44] Startups such as Moon Juice and Your Super, both located in California, offer superfood powder mixes that contain spirulina and chlorella for making smoothies, juices, or tossing into your morning cup of joe or baked goods for a shot of micronutrients.

Big food companies are investing in the research and development of microalgae protein, which ideally should bring the cost of production down, and make this plentiful protein source available to all. For example,

Unilever has partnered with UK-based biotech startup Algenuity to bring microalgae into future foods for Unilever's plant-based portfolio. Algenuity works with a specialized, nutrient-rich microalgae called *Chlorella vulgaris*, which the company has been able to improve for mainstream use by removing the bitter notes and vibrant green color. The partnership with Unilever's Foods and Refreshment division will explore ways of bringing foods made with microalgae to the market.

We have a lot to look forward to as companies and biotopian chefs dream up new foods and dishes using this abundant, climate-friendly micro-food source that

The FUL Story: From Micro(algae) to Macro(impact)

FUL is derived from the tiny but mighty microalgae—the superheroes of the natural world. Microalgae have been powering the earth for over 2 billion years and gave rise to life as we know it on this planet by producing over half the oxygen we breathe.

Inspiring a planet positive future: Microalgae are more effective than trees in transforming CO_2 into an all-natural source of vitamins, minerals, and proteins—offering a more sustainable way to nourish ourselves and our planet. Microalgae require no arable land to grow, no chemical pollutants, and can result in no biodiversity loss. They even grow vertically!

FUL of life: FUL is packed with key vitamins, minerals, antioxidants, and plant-based proteins.

The power of small: Inspired by the big role small actions can play in benefiting our planet and our communities, the three founders launched FUL Foods in pursuit of a planet-positive future.

—FUL Foods[45]

may be the lowest on the food chain but packed with superior nutrient powers.

Microalgae offer much untapped potential as a viable, climate-friendly protein alternative. They have a significant part to play in food system transformation.

—Alejandro Amezquita, future bio-based ingredients R&D director, Unilever F&R

Living Architecture

Your next home just may be built by bacteria. At least that is the hope of a team of researchers from the University of Colorado Boulder who are working with mighty micro-contractors.

Some of the materials in our buildings and home are biologically sourced from nature, such as wood, but it is no longer alive, having been cut and treated. But engineer Wil Srubar, an assistant professor in the Department of Civil, Environmental and Architectural Engineering (CEAE), and his colleagues questioned why we couldn't keep material alive. Their quest led them to experiment with cyanobacteria, green microbes

The Biotopian Chef

Spanish chef Ángel León, of the two-Michelin-star restaurant Aponiente, near Cadiz, Spain, has been cooking with plankton—microalgae that he catches with his own nets in the ocean. After five hours the nets only return one gram of the microscopic organisms, which León farms in tanks, where the plankton reproduce and multiply. He then freeze-dries the plankton into a fine matcha-like green powder. He reconstitutes the dried plankton in water, making dishes like risotto, or cocktails, giving the culinary sensations an oceanic saltiness.[46]

that, under the right conditions, absorb carbon dioxide gas to help them grow and make calcium carbonate—the main ingredient in limestone and cement.

The team is developing "bacteria bricks" that they manufacture by injecting colonies of cyanobacteria into a solution of sand and gelatin. Prompted with the right tweaks, the microbes create a calcium carbonate that mineralizes the gelatin, which binds together the sand.

In their study, the researchers found that the bacteria-churned bricks have the same strength as the mortar used by contractors today, and they held up under a range of humidity conditions. The researchers also discovered that these bioengineered bricks may be super-sustainable in that they could reproduce more bricks by just chopping one brick in half.

Bacteria have been tested in concrete by prior researchers to aid in self-healing, but the challenge so far has been that the survival rate of the bacteria was less than 1 percent, making it not viably self-replicating. Srubar and his team were able to calculate the life expectancy of their bacterial colonies—roughly 9 to 14 percent were still alive in their materials after thirty days and three different generations in brick form. The carbon dioxide lifeline of these embedded bacterial bricks earns them another gold star as they can also remove carbon dioxide from the air.

Before you start racing to your local Lowe's, there is still much work to be done to make bacteria-produced bricks commercially viable. For example, the cyanobacteria used needs humid conditions to survive, and that would rule out more arid regions of the world. But fear not, the team is already working on engineering microbes that resist drying out and remain alive and functional.[47]

Just Add Water

Once these microbes are successfully bioengineered, there is a huge potential that you and I, or your local contractor, could build the microbial home of your dreams by just adding water to bacteria-injected sand and watching it grow. Most of our building materials may start with just sacks of bacterial sand that are easy to ship. We may finally realize low-cost, climate-friendly housing that can self-repair, breathe, regenerate itself, and clean the carbon dioxide out of the air around it.

The possibilities for engineered living materials are endless. Researchers have already demonstrated electrically conductive biofilms, living photovoltaics producing a small electric current from bacteria, and living masks that senses and communicates exposure to toxic chemicals. Applying these in scale to our building materials will help tackle the challenges of climate change, disaster resilience, and aging and overburdened infrastructure.[48] The future of our built environment will be alive, resilient, and teeming with our friendly bioengineered microbes. We can't wait to get building our microbial home, can you?

Biocement

Biomason, a biotechnology company, grows its precast materials by employing microorganisms, similar to the process of hydroponics and traditional concrete block manufacturing. They mix waste aggregate with bM natural bacteria and a solution of water and a calcium source to create a material that hardens to full strength in less than seventy-two hours, with the same composition as natural stone.[49]

The energy-intensive processes of making cement range from extracting the raw material, transportation, and fuel sources for heating kilns, and these processes contribute to the fact that "40 percent of global carbon dioxide emissions are linked to the construction industry."[50]

In Living Color

There's another feature we don't think about in our average bacterium, and that is its vibrant color. In a research collaboration between the University of Cambridge and Hoekmine BV, a Dutch company, a team has shown that we can change the color and appearance of certain types of brightly colored bacteria by altering their genes. The bacteria studied were *Flavobacteria*, commensal bacteria found in soil and fresh water that produce vivid metallic colors, which from their internal colony structure reflect light at certain wavelengths.

The study identified the genes for structural color in nature, like a butterfly's wings and a peacock's feathers, that produce colors with microscopically structured surfaces fine enough to interfere with visible light—selectively absorbing wavelengths. Take a peacock, for example: its tail feathers are brown, but their microscopic structure makes them also reflect blue, turquoise, and green light—often with an iridescence.

By genetically mutating the bacteria, the researchers were able to change the geometry of the *Flavobacterium* colonies, which changed their colony color from the original metallic green color to blue and red. The team was also able to manipulate the vibrancy of the color, making it duller and even eradicating it entirely.[51]

The potential is huge in using this genetically altered bacteria as a tunable, living photonic structure that can be reproduced in abundance, whose colors can change when triggered by external stimuli. We can just imagine when biodesigners get their hands on these living "photoshoppers" that can change color with a slight tweak. We can one day have biodegradable, easily changeable living paints for our walls, cars, and other products, simply by growing the exact color we want.

The Shades of Bioleather

Material research and design Studio Lionne van Deursen created leatherlike sheets of bacterial cellulose from yeast and bacteria, with every sheet having a different color and translucency. They used the biomaterial in their Imperfect Perfection lamp to diffuse the light source with the unique hues of each sheet produced.[52]

B Is for Bacteria

Biodesigners can never start too young! As a part of the push for STEM (science, technology, engineering, and mathematics) learning, parents, authors, and educators are eager to stimulate the imaginations of the next generation of biotopians to foster curiosity about and, just maybe, new love for bacteria by portraying them as something other than icky germs. It's not often that bacteria get to be the protagonists, but in the book *The Squid, the Vibrio, and the Moon* the heroes are bacteria that help a young bobtail squid evade its predators. The setting for the book *Zobi and the Zoox* is in a coral colony on the Great Barrier Reef, where the protagonist is a rhizobia bacterium—Zobi for short. Both of these books give children a greater appreciation for the many roles that bacteria play in making the biosphere work. It is easy to say that bacteria are an important part of all ecosystems, but that statement needs to be followed with great examples of actual symbioses like these books provide. Another great addition for studies of the microbial world is *Inside Your Insides: A Guide to the Microbes*

That Call You Home. This book has a wide range of information about microbes—what they are, where they live, and what they have to do with us and our world.[53]

Magnificent Microbes

As the microbial world is becoming more visible, we are coming around to appreciate the elegance of its design. The growing need for cross-disciplinary collaboration and creativity prompted by scientific research will continue to expand our design and manufacturing capabilities. We are witnessing today the rise of homegrown biology, much like the "maker" movement in robotics. This coming together of experts, amateurs, designers, scientists, the old and the young will create a new platform for innovation.

Several exhibits have brought the microscopic and invisible world of microbes to life. *Community of Microbes* lets you step in virtual fun with an augmented reality that immerses you in a colorful, sensorial experience that trans-

Living Patterns

Synthetic biologist Tal Danino manipulates microorganisms in his lab to create arresting, colorful patterns—and to fight cancer.

In his day job, leading the Synthetic Biological Systems Lab at New York City's Columbia University, Danino and his team are engineering bacteria—*E.coli*, an *E.coli* probiotic, and Salmonella—to detect and treat cancer. Yet, in 2015, he also programmed bacteria's unique properties to make living art for an exhibition at MIT. Danino has a few collaborative art projects with bacteria: working with conceptual artist Anicka Yi, he created an installation of bacterial cultures taken from the bodies of one hundred women. He also collaborated with the artist and photographer Vik Muniz on a series of ceramic dishes inspired by bacterial patterns.

In *Microuniverse*, Danino used different species of bacteria grown under different conditions and time lengths, creating a series of abstract images. He manipulated the conditions of temperature and humidity under which the different species were grown. Danino used about twenty different species of bacteria for *Microuniverse*, letting them grow for as few as two days and as long as two months. He partnered with the company Print All Over Me to create custom apparel based on the images of bacteria from *Microuniverse*, with part of the proceeds going to cancer research.[54]

Amino Labs

Amino Labs democratizes access to learning and innovation in the world of biotechnology. One of its tools is its Engineer-it Kit for genetic engineering at home or in the classroom. The kit contains all ingredients needed to grow and engineer bacteria with DNA to create color pigments. Its all-in-one beginner kits let you paint with colorful bacteria; engineer living colors, smells, and enzymes; control genetic circuits; and more—without any prior knowledge or skills.

Amino Labs' mission is to make biotechnology learning and innovation accessible, and to discover how most medicine, food, and modern materials are made. Students and families can learn how to genetically engineer microorganisms, create living bio-art, and grow and extract bio-products.

They believe that making the tools for "programming biology" accessible to non-scientists will be key in driving a "wide-spread meaningful growth of ideas and solutions to our greatest problems." For the kids in all of us (aged twelve and older).

—Amino Labs[55]

formed the Cooper Union Colonnade at Cooper Union in New York City into a microbial wonderland. Created by multidisciplinary artist Amanda Phingbodhipakkiya in partnership with microbiologist Dr. Anne Madden, the exhibit takes you into the world of the diverse and vibrant ecosystems of microscopic organisms that live in, on, and all around us—the microbes in our dust that create pharmaceuticals, to microbes that make squid glow and beer taste good. The project was brought to life with the support of a diverse group of science advisors, designers, animators, developers, science writers, and STEM advocates.[56]

Our Biofermented Future

Look around you—almost everything you wear, decorate your home with, eat or drink, put on your skin, renovate, or build with may one day be made through biofermentation. Biofermentation is the technology of the growing global bioeconomy, the new tech gold rush that McKinsey reports will become a $4 trillion industry.[57] From foods to textiles to medicines, synthetic biology holds the promise of high-quality, sustainable ingredients made with biofermentation instead of being extracted from natural resources.

As we've learned, microorganisms are the new design tools, biology their template, and microbial biofactories our manufacturing and production facilities. Microbial "biofacturing" uses reactions of microbes (like bacteria or yeasts) to make the product we want, engineering a synthetic process that is smarter than a living cell.

The challenge in the coming years is scalability—expansion that will generate greater output and allow wider access to biofacturing. When we scale up new technologies like biofacturing, it ultimately reduces production costs and increases profitability as the biomanufacturer, bioengineer, or biodesigner is able to achieve growth, by producing more.

One challenge in scalability is the timing. In mechanical factories, you just turn on and turn off machines that do the work for you. Some may break down, but are repaired or replaced, and any well-oiled factory has backup to

Building Better Together

BioCurious

Meet up and get your microbial design powers on! BioCurious is building a community biology lab for amateurs, inventors, entrepreneurs, and anyone who wants to experiment and learn to use biotechniques. Located in Santa Clara, California, the community lab will also be a meeting place for citizen scientists, hobbyists, activists, and students.

—BioCurious[58]

Genspace

Since 2009, Genspace, the world's first community lab, in Brooklyn, New York, has been the place where anyone can learn and work on biotechnology. A nonprofit organization, it provides STEM educational outreach, adult classes, cultural events, and a platform for science innovation at the grassroots level. The best way to inform twenty-first-century dialogue about science is to have stakeholders understand it from a hands-on perspective. Genspace fosters a safe and inclusive community where all people— including those from nontraditional and underrepresented backgrounds—can experientially learn and boldly create with life sciences.

—Genspace[59]

Grow Your Ideas
iGEM

The iGEM Foundation, a nonprofit organization dedicated to the advancement and education of synthetic biology, runs the iGEM Competition that gives students the opportunity to push the boundaries of synthetic biology by tackling everyday issues facing the world. Projects from iGEM serve as proofs of concepts and prototypes, contributing to over 150 startups.

—iGEM[60]

OpenCell

OpenCell offers private and shared laboratories to rent—fashioned out of seventy shipping containers in Shepherd's Bush, London. OpenCell transformed these containers into affordable biolaboratories for biotechnology startups, offering a low-cost environment for biotech prototyping.

—OpenCell[61]

ensure it keeps producing every minute. Microbial biofactories can't operate nonstop (no living organism can). The characteristics of microbes make them less of a churning machine and more of a "working population" that over a short amount of time become less productive—so, naturally, scientists are looking at ways to make microbial biofactory workers more efficient and more productive.[62]

AsimicA, a biotech startup, is working on a new technology they call "microbial stem cell technology" or MiST that they think will raise the yields in biomanufacturing, and keep the bioreactors running longer. A spinoff from a research project at the University of Wyoming, Laramie, AsimicA's method helps to repopulate the biofactory with young and productive microbes during the fermentation batch. Much like how human stem cells divide and grow into newer cells, MiST delivers stem cell-like properties to microbes, changing how they grow, divide, and self-renew. Normally, microbes divide into two identical twins but microbial stem cells divide asymmetrically, renewing themselves while also generating new cells, bringing a younger and more productive workforce to the microbial biofactory.[63]

There is great hope that by educating and providing the tools for biodesigning and biomanufacturing with microbes to students, artists, entrepreneurs, and technologists, we will finally create new proteins, new foods, and new fashion materials without killing animals or plants. We will design new targeted treatments for the benefit of human, animal, and agricultural health that may even stop the risk of disease. And we will be able to do this all faster and with environmental responsibility. It turns out that biology is our new humanitarian technology, and once we unleash the industrious nature of microorganisms and their counterparts in nature, we will experience exponential advances where, as the great geneticist George Church explains, we can bring down the costs of production while raising everyone's standard of living. Imagine that!

Certainly, some engineer or company is going to try, and in many cases succeed, to make almost every material biologically. And by materials, I mean replacements for concrete, building materials, metals, electronic components, hard-to-manufacture materials that have very specific shapes that are hard to make by bulk processes. Possibly carbon nanotubes, which have interesting electrical conduction and mechanical properties. Things like synthetic diamonds.

There's nothing that *a priori* at this point, it is impossible for a biological system to make, and make it in a way that's greener, less energy consumptive, more precise. Biology can make very complicated materials like, say, wood, which at the atomic level is precise. We may not use all that complicated, intricate components when we burn it or carve it or use it as a material, but it's there. And you can make it very, very inexpensively, on the order of a fraction of a dollar per kilogram. Every time you can bring down costs, there's a potential for that to raise the standard of living. If we can reduce the cost of building basic infrastructure like clean water and roads and providing transportation, that's going to have a transformative effect on developing nations that are already, some of them, coming up out of poverty.

—George Church, geneticist pioneering the Human Genome Project and the Personal Genome Project, interview with Sputnik Futures, 2011

Ancient High Technology

Harnessing MI (Microbial Intelligence)

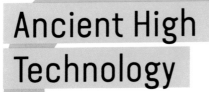

Where Do They Get All Those Wonderful Toys?

Super Microbes Can Swallow Pollution, Create Energy, Smash Waste, Heal Humans . . .

The story has been told . . . we need to end our reliance on fossil fuels. Meat is on our high carbon footprint menu. Fires, floods, famines are on the rise. Disruptive climate change is the lead story on our favorite newsfeeds. Fast fashion, waste, and plastics are swallowing our land and oceans. Superbugs are overtaking our antibiotics. Face it—we're living in a dark, Gotham-like world of environmental corruption. Thank goodness Commissioner Gordon is flashing the spotlight. Help is on the way. Promising research that we might be able to pull carbon dioxide (CO_2) out of the atmosphere and slow the pace of climate change may soon be a reality.[1] It's not *impossible* to grow a burger (hold the pickle, hold the lettuce, hold the high carbon footprint). We have the means and the ability to grow high-quality protein, clean up our pollution, and face the perils of antibiotic resistance. The superhero commandeering these huge breakthroughs: bacteria. This tiny yet mighty microbial force sweeping our Gotham cloud has the future of our health, wealth, security, and lunch in its elegant superpowers. Where do they get all those wonderful toys? Let's unmask bacteria's secrets now . . .

Microbial intelligence (more popularly known as bacterial intelligence) is characterized by the uncanny, adaptive behavior shown by microorganisms. In other words . . . bacteria are far more street-smart than we realize!

As we've seen, bacteria adopt

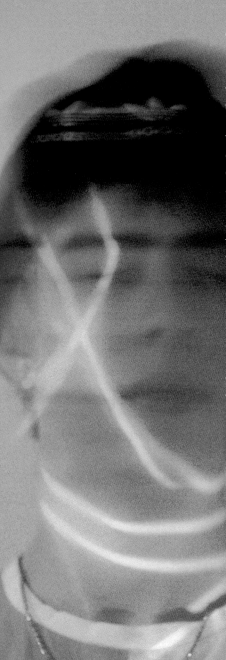

different survival strategies to make their life comfortable. Many bacteria, including *E. coli*, are capable of communicating among themselves to determine if microbes nearby are friend or foe through "quorum sensing"; others bond together to build communities, search for food, and influence survival instincts. Now that we see how bacteria are, in their own special way, big thinkers, the more we know how they "feel" about their surrounding environment, the better we can investigate different ways to design—and build—with them. Microbial intelligence (MI) is ancient high technology, and harnessing it could lead to innovations in diverse fields from food to fashion to architecture, to medicine and agriculture.

We're discovering how bacteria can "eat" pollution and waste and create energy—just to name a few of their superpowers. We have so much yet to learn from our industrious micro-ancestors. For over 100 million years bacteria have been talking—and today we're starting to listen and learn!

Bacteria Talk Shop

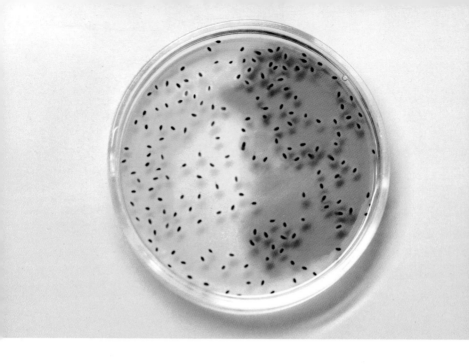

Archaea: Hyper-Resilient Microorganisms

The Bioengineer's Muse: 100 Million-Year-Old Miracle Microbes

Bioimagine this: the first designers and engineers on planet Earth could still be here, surviving millions of years sequestered in the deepest, darkest parts of the ocean floor, the depths of hot springs or glacial basins, and the subterranean strata, surviving in the most unwelcoming environments.

Now, bioimagine a beautiful collaboration where we awaken these microbial survivors and start to tap their regenerative survival skills—perhaps, we could one day solve some of the world's most intractable ecological problems. We are not far from this possibility.

The ever-expanding need for sustainable solutions is driving remarkable research based on biological ecosystems. Since the first microbial matter was extracted from hot springs in Yellowstone National Park in 1966 in the hope to understand how life can survive extreme heat, scientists have been discovering a variety of microbes in desolate deep seabeds that have upended conventional wisdom about the requirements for sustaining life. Scientists are learning that within rocks and inside frozen, toxic, and extremely hot places, life not only exists but thrives. For example, one such microorganism found deep below the surface has been buried for millions of years and may not need to rely at all on energy from the sun. Instead, the microorganism has found a way to create methane in this low energy environment, which it may not use to reproduce or divide, but to replace or repair broken parts.[2] The continual discovery of microbes in the most extreme environments on Earth has led to the addition of a new branch on the tree of life that encompasses these hyper-resilient microorganisms called archaea. The potential for bioengineers and scientists to develop new and useful applications with this previously hidden and unseen form of life is game changing.

One of the first major discoveries was a bacterium uncovered in 1966 called *Thermus aquaticus*. The hot springs of Yellowstone National Park are known for their vibrant colors, which come from microorganisms like *T. aquaticus*, which thrive at a scalding 160 degrees Fahrenheit, about 30 degrees hotter than what was thought to be the limit for life at the time. Because of its ancient age and tendency to live in extremely high temperatures, *T. aquaticus* may be a holdover from the dawn of early Earth, thought to be a hot and steamy environment.[3]

This ancient, heat-loving microorganism has had perhaps the largest impact on DNA research and the Human Genome Project, all because its proteins still functioned at 130°F, where those of most cells did not. Proteins are what is needed to make copies of DNA, and are the key to Polymerase Chain Reaction, or PCR, the method used to extract the DNA information out of our cells. PCR requires heating the DNA to extreme temperatures, which destroys the proteins needed to copy the DNA.[4] But along came *T. aquaticus* with the enviable

protein that withstood temperatures above the threshold of most proteins, surviving where others failed. From this organism was isolated Taq DNA polymerase, a heat-resistant enzyme crucial for a DNA-amplification technique widely used in research and medical diagnostics. Think of it as a photocopier. *T. aquaticus* has become an essential tool in almost every biotech lab around the world.

In fact, we're relying on *T. aquaticus* (the "hot" bacteria) found in the thermal lakes of Yellowstone to be a key component in most COVID-19 testing. Testing for many viruses, including coronavirus, uses the PCR technique (repeatedly heating and cooling a sample), like that from a COVID-19 nose swab. During the process, the DNA in a given sample doubles after each heating and cooling cycle, making it easier to see what kinds of DNA there are. It's a tricky process, though, because if genetic material gets too hot, the strands essentially dissolve and become useless. Which is why you use the enzyme from *T. aquaticus*—its heat-resistant superpowers protect the genetic material from being destroyed by the heat. PCR has been around since 1983, and the latest application in coronavirus testing has helped to enable the vital rapid results.[5] However, it's not as simple as it seems. The genetic material in the novel coronavirus is RNA, a single-stranded cousin of DNA. PCR can only be used with DNA, so to use the PCR in the testing, scientists first convert the RNA to DNA.[6]

After Dr. Thomas Brock and his colleagues discovered *T. aquaticus* in the late 1960s, the enzyme wasn't identified until the 1980s when Dr. Kary Mullis first isolated the Taq polymerase, which he used to develop the Polymerase Chain Reaction (PCR) technique that enables scientists worldwide to replicate millions of copies of a DNA sequence—and garnered Dr. Mullis the 1993 Nobel Prize in Chemistry. Today, PCR is regularly used in medical research, genome mapping projects, and even crime scene investigations. And the laboratory that started it all—the Lower Geyser Basin hot spring of Yellowstone—continues to be a research habitat, with 25 percent of the Yellowstone research permits issued annually to investigate the unique microbial communities thriving in its hydrothermal wonders.[7]

Munching on Gas, Twenty Thousand Leagues under the Sea

Researchers from the Max Planck Institute for Marine Microbiology and MARUM, the Center for Marine Environmental Sciences at Bremen University, have discovered *Ethanoperedens*, microbes that feed on ethane at deep-sea hot vents. The team of researchers have been able to cultivate this microbe in the laboratory and found that the mechanism the microbe uses to break down ethane is reversible. In other words, it is possible *Ethanoperedens* can both consume ethane and produce it.

Ethane is the second most common component of natural gas, comprising about 15 percent of its chemical makeup, so just imagine the possibilities if *Ethanoperedens* could produce ethane as a renewable energy source. This is the goal of a massive body of research in developing biomass fuels for transportation and electricity, thereby moving us toward clean and renewable energy sources.

Biomass fuels (i.e., fuels made from renewable organic material derived from plants, animals, and now microbes) are gaining traction in developed markets, and one day our microbial friends may be our most critical partners for decreasing our dependence on fossil fuels and climate-threatening carbon dioxide emissions.[8] But breaking down natural gas requires microbial teamwork.

While some bacteria can break down the natural gas components propane and butane, they alone can't degrade the bulk of natural gas—methane and ethane. What's needed is a consortium of archaea and bacteria, whose biochemical processes work in tandem to break down natural gas. So far, unfortunately, their productivity has been very slow. The microbes eventually run out of steam, and since they only divide into new, stronger workers every few months, bioengineers face many of the same challenges as the biodesigners we met in chapter 4. This helps to explain why there hasn't been a mass quantity of biomass fuels yet.

Still, there is hope: the researchers from Bremen with their discovered archaea named *Etha-noperedens thermophilum*—which means "heat-loving ethane-eater"—are reporting that this consortium is growing much faster, with the cells doubling every week. More research is needed, and the team from Bremen and its research colleagues are still searching for other microbial cooperative workers that may be dwelling in the consortium, to adopt their natural cycles for developing biological solutions for reducing carbon emissions.[9]

The Ice-Microbe Cometh

Permafrost could become a gold mine for humans. On one hand, thawing permafrost can destroy homes and roads (a tragedy already reported in parts of Alaska) and release tons of carbon dioxide into the atmosphere. On the other hand, the microbial diversity and organic compounds that have been frozen for tens or hundreds of thousands of years may have some amazing things to teach us. The changing Earth's climate is causing permafrost to melt all over the world, revealing interesting species of bacteria that scientists hope can result in new environmentally friendly solutions.

Researchers at the Swiss Federal Institute for Forest, Snow, and Landscape Research (WSL) gathered samples from the Schafberg mountain three thousand meters above Pontresina (eastern Switzerland), where the permafrost is estimated to be thirteen thousand years old. The sample they gathered from the permafrost soil had much greater microbial diversity than the soil found on the surface. The team already found another new type of yeast that thrives at -5°C, and they speculate that they may discover over one hundred new species. What the researchers are most interested in are the hypothermic species—the microbes that can endure extreme freezing temperatures.[10]

Most mammals, including humans, experience hypothermia when the body loses heat faster than it can produce heat. But fans of space travel films such as *Interstellar* will recognize it as the suspended "long sleep" astronauts placed themselves in to physically endure the stress of time and the impact of microgravity on the body. Well, hypothermic suspended animation is moving from sci-fi to a possible solution for deep-space travel, as researchers are now considering the medical practice of therapeutic hypothermia.[11] Interestingly, this practice dates back to around 400 BCE in the writings of Hippocrates, who mentions the use of snow and ice to reduce hemorrhage in patients.[12] Exposure to extreme cold temperatures has been shown to increase metabolic rate and stimulate the generation of brown fat—the fat tissue that helps the body to burn more calories to keep warm, which burns off excess body fat. Ice therapies have gained popularity lately for immune boosting, reducing stress, inflammation, and fatigue, and the quick recovery of sore muscles.[13] Ah, perhaps the hypothermic species of bacteria being uncovered in the melting permafrost will accompany the first mission, manned or unmanned, to Mars (but more on microbes and space travel in chapter 6).

The research findings in hypothermic microbes have already yielded new products and po-

tential solutions, one example being enzymes that can break down fats at lower temperatures, which could be used in detergents for better cold-water wash cycles that would help conserve energy. Researchers are also investigating the possibility that some of these bacteria could be employed to break down plastic, potentially one day putting them to work in commercial-size bioreactors that could replace traditional recycling centers by having the microbes eat away at the plastic rubbish.[14]

When It Comes to Climate Change, Root for the MVP!

Thirty microbiologists hailing from nine countries published an alarming but rallying call in the June 2019 issue of the journal *Nature Reviews Microbiology* warning humanity to stop ignoring the Earth's MVP (most valuable player): microbes. The researchers want to encourage their colleagues to include microbes in climate change research, increasing the use of research involving innovative technologies, and improving education in classrooms. The modest yet unseen microbial superstars are, in fact, the critical players in the Earth's biodiversity and ecosystem when addressing climate change. Microorganisms, which include bacteria and viruses, are the "unseen majority" of life-forms on Earth, and enable vital functions in animal and human health, agriculture, the global food web, and industry.[15]

Take the marine life food web, for example. According to the Census of Marine Life, 90 percent of the ocean's estimated total biomass is microbial. In our oceans, marine life-forms called "phytoplankton" work just as hard as plants do for Mother Earth, taking light energy from the sun and eating and removing carbon dioxide from the atmosphere. Tiny phytoplankton form the beginning of the ocean food web, feeding krill populations that then feed fish, sea birds, and large mammals such as whales.[16]

Bacteria's MVP status comes from their innate skill and ability to "eat" CO_2, our principal greenhouse gas. The only drawback is bacteria eat CO_2 slowly. But, for better trained microbes, we just need to look at the work of the Weizmann Institute of Science, Israel, that is pumping out faster-growing microbe species that could remove more CO_2, by modifying E. coli through the insertion of DNA from CO_2-consuming photosynthetic bacteria. Their work provides a proof-of-concept that dramatic changes in the metabolism of commonly used microbes are attainable. In theory, a solar panel-E. coli system could be ten times more efficient than photosynthesis at removing CO_2 from the atmosphere. As the research develops, it could have an important positive impact on global warming.[17]

Back on land, researchers from Cornell University's School of Integrative Plant Science in Ithaca, New York, are experimenting with the dirt on our playing field with carefully selected microbes that both boost the soil's carbon-storage capacity and accelerate the growth rate of crops planted in it. A large amount of carbon is absorbed by soil—about 25 percent of the world's fossil fuel emissions are removed by soil each year—and small changes in

how we manage soil could make a big impact on climate change.[18] Understanding how bacteria break down carbon in soil could hold the key to the sustainability of soil and the ability to predict the future of global climate.[19] So, when it comes to climate change, if our MVPs—microorganisms— aren't considered effectively, it means climate models cannot be generated properly and we will be missing a very big component of the equation.

Bacteria Buffet . . .
The Big Cleanup

Pull up a chair—researchers around the world are inviting bacteria to the plastic buffet. Bacteria are the ultimate dinner guests and are useful diners at wastewater treatment plants. After all, sewage is like a fine wine for bacteria. Bacteria are also gluttons for carbon, and some microbes will even mutate to slurp up oil spills and other forms of pollution. Scientists are analyzing bacterial communities in soil near oil spills to identify which hungry species would be best at degrading the pollution. Biosimulation is an approach that scientists are using to help bacteria consume their food faster. The idea is that the right oil-eating microbial mixture could be injected into the dirt— bon appetit!

For some bacteria, plastic is the main meal! This is especially good news since plastic is one of the biggest pollutants and does not degrade easily—an average mineral water bottle takes half a millennium to decompose.[20] Researchers at the Ben Gurion Institute of Science, Israel, have developed bacteria called *Pseudomonas putida*, which can be genetically modified to eat polyethylene-terephthalate (PET), the most common kind of plastic polluting the oceans and land.[21]

The bacterium uses plastic as a carbon source and breaks it down using engineered enzymes into smaller molecules such as ethylene glycol, which can be recycled to make new plastic. This could be a small answer to the world's very big plastic problem.[22]

PET isn't the only form of plastic bacteria can gobble up either. According to researchers at Helmholtz Centre for Environmental Research, Germany, some soil bacteria will happily snack on polyurethane (commonly used to make insulation and car parts).[23] French startup Carbios has developed a mutated bacterial enzyme that can almost completely digest old plastic bottles in just a few hours—helping turn plastic waste into the chemical block to make new plastics.[24] Also in startup mode, Miranda Wang and Jeanny Yao, cofounders of Novoloop, are turning plastic waste into gold, or more precisely into high-value chemicals. Novoloop takes the dicarboxylic acid it generates from polyethylene and transforms it into high-performance materials, such as photopolymers or polyurethane. Or more simply stated, garbage becomes valuable new feedstock for product development.[25]

Chew on This

Ninety-one percent of all plastics on this planet are not recycled! In a study that did a global analysis of all plastics ever made, the researchers found that of the 8.3 billion metric tons that has been produced, 6.3 billion metric tons has become plastic waste. Of that, only 9 percent has been recycled. And if the trend of plastic waste continues, by 2050, there will be 12 billion metric tons of plastic in landfills . . . which means we need to get trillions and trillions of bacteria to start working on it![26]

From Low-Value Waste to High-Value Molecules

Scindo is developing a biological platform using novel enzymes to recycle the unrecyclables. They are turning the hardest plastics to recycle into high-value molecules, creating a circular alternative to landfills and incineration.[27]

Bacteria Eat Your Dirty Water

Plastic isn't the only culinary delight for bacteria. Wastewater, specifically sewage, is another buffet where bacteria's gorging can be our next generation of recycling by cleaning dirty water. We may think of it as sewage, but bacteria look at it as their buffet. Even just in your home septic tank lives a whole microbial universe of bacteria, yeasts, and enzymes that, together, play an active role in maintaining your septic system. Microbes feast on any solids and get the process of decomposition going.[28] If you have a septic system, here's a trick to keep your microbial sanitation workers happy: once a month, flush a packet of brewer's dry yeast down one toilet on the bottom floor of your house, adding some more "good" workers to your septic tank.

Oxygen-consuming microbes are also critical to wastewater treatment plants, but it's expensive to run on a large scale since the wastewater has to be kept aerated in order to tempt microbes to break down the waste. But one company has a solution to clean industrial wastewater while generating more energy than it consumes: Cambrian Innovation in Massachusetts uses bacteria that can grow in the absence of oxygen, which saves the cost of running air pumps. Their system is like a production line: one cycle is made up of "electroactive" microbes, including *Geobacter* and *Shewanella*, that consume the waste in the water and produce electrons and other charged particles. The other cycle uses some of the charged particles as a source of energy to convert CO_2 into methane gas, which can then be burned to generate heat.[29]

The Dirt on Soil Bacteria

Bacteria are the most populous organism in soil, found in just about every patch of earth, and most of them are not harmful to humans. When was the last time you picked up a handful of soil and smelled its "earthy" scent? What you were inhaling is not a collective scent of the things in the soil, but the chemical release of *Streptomyces*, soil-dwelling bacteria that produce diverse metabolites that impact plant health. But it seems the *Streptomyces* need some help too. Researchers at the Swedish University of Agricultural Sciences in Alnarp found that the odor re-

leased by the bacteria attracts invertebrates that help the bacteria disperse their spores, which are the structures that help them survive adverse environmental conditions. So that sweet, dark earthy smell is, in part, a survival scent of attraction.[30] (Think we have a new perfume idea here?)

Bacteria do more that give us the smell of earth, they perform ecosystem services in the soil, improving soil structure and recycling of nutrients, and are the key ingredients in natural fertilizers. Free-living soil bacteria form relationships with plants, helping them get the proper nitrogen "fix" to help the plant's growth and production. Nitrogen is a major component of chlorophyll, the most important pigment needed for photosynthesis, as well as amino acids, the key building blocks of proteins. But even though nitrogen gas is abundant in the Earth's atmosphere, plants can only use a reduced form of it. That's where soil bacteria like *Azotobacter* come in with biological nitrogen fixation (BNF). These organisms utilize the enzyme nitrogenase to catalyze the conversion of atmospheric nitrogen to the tissue of certain plants to help with development.[31]

Now bioengineers are working on ways to manipulate soil bacteria to perform even greater feats. One way is to help accelerate plant breeding, which plays a central role in increasing crop yield by "breeding" the bacteria that live in soil around the plant. Researchers at Utrecht University have modified the genes of plants' soil bacteria neighbors, a method that can be used in agriculture to improve the health and resilience of crops. It's also faster and cheaper than plant breeding.[32] Reazent, an agtech startup, wants to boost agriculture as well, by supercharging soil bacteria to trigger crops to grow more. A member of the tenth batch of IndieBio's program, Reazent is an organic fertilizer alternative to synthetic agrochemicals.[33]

Microbes can also fight toxic waste in contaminated soils. Allied Microbiota develops advanced microbes to clean up environmental contamination faster, more sustainably, and more cost-effectively than the traditional method of relocating contaminated soil to landfills. A graduate of the IndieBio acceleration program, Allied Microbiota's aim is not to bury the problem but to clean it, turning brownfields (contaminated and unhealthy fields) into vibrant, healthy greenfields.[34]

If everything we've touched on isn't enough to convince you to invite bacteria to your next soirée, consider this: the future can embrace an abundant circular economy that turns wasted carbon, plastics, and more into high-performing materials that can be recycled over and over again. Think of it as using more biological resources to help use fewer natural resources. Now that flips the scarcity model!

Bacteria's Energetic Future: Hint . . . It's Renewable!

Traditional biofuels are produced from food crops, such as wheat, corn, soybeans, or sugarcane (the so-called energy crops).[35] This approach is costly and competes with food production in the use of land, water, energy, and other environmental resources. In order to make biofuels a net positive for climate change and sustainability, we have to do it in a way that doesn't compete with food production. Using the food supply is a zero-sum game, and if you're relying on the same sources, anything you do to make biofuels will be interfering with agricultural production. But, perhaps, there is another way. If you guessed it has to do with bacteria, you're getting the gist of this chapter. Bacteria are extraordinary chemists that can churn out renewable biofuels. Let's look at the work that is in progress today.

The most common biofuel in the United States is ethanol, mostly made from corn starch, which means

Drying Out Microbial Diversity

A teaspoon of soil contains up to a billion bacteria, yet, like many plant and animal communities, soil microbes may be facing new threats due to climate change—specifically the increase of dry lands. Dryland ecosystems cover about 41 percent of the Earth already. Little is known about whether this increasing soil aridity will cause a loss of microbial diversity. One research group found that a loss in microbial diversity would negatively impact the key ecosystem functions in soils vital to global food production.[36]

From Biomass to Biogas

Biomass was once the largest source of energy in the United States until the mid-1800s. It continues to be a source for fuel in developing countries, mainly for cooking and heating.

One of the cleaner ways to convert biomass to energy is through biological conversion, which includes both fermentation to convert biomass into ethanol and anaerobic digestion to produce renewable natural gas—also called "biogas."

In 2019, biomass provided nearly 5 percent of total primary energy use in the United States. Of that 5 percent, about 46 percent was from wood and wood-derived biomass, 45 percent was from biofuels (mainly ethanol), and 9 percent was from the biomass in municipal waste. Transportation and industry accounted for the largest percentage of total annual US biomass consumption.

—US Energy Information Administration [37]

it competes directly with the food supply. Ethanol is also somewhat volatile and has a lower energy content than gasoline, making it a less-than-ideal alternative fuel. The isobutanol that Jason Peters, an assistant professor in the University of Wisconsin–Madison School of Pharmacy's Pharmaceutical Sciences Division, and his team are producing will be made from inedible plant material, will be less volatile, and will be a more direct replacement for gasoline. Their bio-innovation will not only provide a cleaner and less volatile biofuel to gas up the world, it will help curb the top threat on WHO's (World Health Organization) list—climate change. [38]

Another unique approach to the challenge of making and scaling biofuels is the use of gene-modifying technology. Peters and his team are using the CRISPR gene-editing technique to reengineer bacteria, such as *Zymomonas mobilis*, to produce isobutanol, a less volatile and higher-energy alternative to ethanol as a replacement for gasoline. [39]

Making Electricity Out of Air . . . and Microbes

There are several other ways in which bacteria can generate biofuels, and the next idea is electrifying . . . literally. Did you know that nature has its own power grid wired from bacteria? Electroactive bacteria have been humming long before Thomas Edison flipped the switch. One example involves a species of bacteria that eats and excretes electrons (negatively charged subatomic particles), generating electricity. Stick an electrode in the bed of a hot spring and watch a colony of these electroactive bacteria grow around it.[40]

Derek Lovley, a microbiologist, and Jun Yao, an electrical engineer, both at the University of Massachusetts Amherst, have spent several years working out how to tap these electrifying microbes for renewable energy generation. Lovley calls the process "microbial electrosynthesis," and it carries the potential to generate biofuels more efficiently than agriculture can. To that end, Lovley and his col-

Algal Fuels

There is an increased interest in the industrial cultivation of microalgae to produce biofuels. Microalgae-based fuels are ecofriendly, nontoxic, and with strong potential of fixing global CO_2. Microalgae feedstock does not require farmable land and fresh water for cultivation and aren't edible, having no effect on our food chain. Asia, Europe, and the United States have started industrialization of bioenergy from microalgae biomass.[41]

leagues have introduced the Air-gen device, which harnesses the tiny electrically conductive hairs of electroactive bacteria in order to generate electricity. The Air-gen has significant advantages over other forms of renewable energy including solar and wind because unlike these other renewable energy sources, the Air-gen does not require sunlight or wind, and it even works indoors.

Their Air-gen device pulls electricity directly from the air, apparently taking advantage of differences in humidity in order to do so. Lovley and his team calculated that they could scale up Air-gen devices to outperform today's commercial solar panels—and unlike solar panels, Air-gen would work at night too.[42]

Bio-batteries could also be on the horizon. A group of scientists are looking at the process of bacterial respiration, which is the process by which bacteria convert the nutrients they take in into power for the biochemical processes they carry out. Typically, respiration in microbes generates excess electrons that are passed to oxygen, but for microbes that live in oxygen-free mud sentiments, the electrons are passed to other mineral ions. A range of microorganisms have developed ways of directly exporting the excess electrons from the cell, shuttling the electrons along tiny molecular wires that protrude from the cell surface. Think of it like a group of spores playing a game of tag.

Research by a team at the Department of Biological Sciences, University of East Anglia in Norwich, England, have shown how thousands of tiny molecular wires embedded in the surface of a bacterium called *Shewanella oneidensis* can directly transmit an electric current to inorganic minerals such as iron and manganese oxides, or the surface of electrodes. Grow these bacteria on an electrode and you end up with the anode (positively charged) half of a battery. Couple this to the cathode (negatively charged) half and feed the bacteria with carbon-based organic matter from industrial wastewater, and you can basically construct a microbial fuel cell that generates small amounts of electricity.[43]

Geobacter sulfurreducens is another bacterium that produces nanowires, protein filaments with metallic-like conductivity that facili-

tate long-range electron transport, which has important implications for the emerging field of bio-electronics.[44] Perhaps one day we will replace the homemade potato clock, where the spud's acid reacts with a positive and negative electrode to create a small electric current, with the nanowires of *Shewanella* or *Geobacter* bacteria . . . growing in a petri dish on your bedside table.

Bacteria, the Original Programmers

In many ways, bacteria are helping us to survive, but it is their own survival technique that has unlocked the future of our life alongside microbes. It is bacteria's survival technique—a technique it has used for millions of years, but that we only discovered in 2012—that may change our ability to program the DNA of life forever. Called CRISPR-Cas9, this high technology uses a smart Trojan Horse strategy where bacteria store genetic clippings that they "steal" from invading viruses, safekeeping their enemy's clippings within their own DNA. That way, when the next viral attack happens, the bacteria recognize their assailant and release proteins to "slice out" the offending virus's DNA at the precise locations identified by—that's right—the clippings they stored earlier.[45]

Researchers are harnessing bacteria's age-old tool of selectively deleting specific genes for some of our greatest health challenges today—from treating cancer and incurable diseases to improving longevity.

We still have a lot to learn from bacteria's clever editing of their DNA, and how to put their smarts to work for us. One study is looking at how bacteria can be programmed to behave like computers, inputting commands and code that act like instructions into their genes, causing them to assemble themselves in shapes and networks. The key is triggering them to communicate, like the internet protocol that allowed, and eventually triggered, computers to talk to each other and other servers, guiding the communication path of billions of devices. Well, researchers did the same thing to bacteria, allowing them communicate with each other, with millions of them gathering in a predictable manner.

The Problem-Solving Amoeba

Japanese researchers from Keio University, in Minato, Tokyo, have demonstrated that an amoeba, a single-celled organism, was able to find a somewhat optimal solution to the question: *Given a list of cities and the distances between each pair of cities, what is the shortest possible route a salesperson could take that visits each city only once and returns to the origin city?* This is known as the Traveling Salesman Problem (TSP), an NP-hard (problems that are informally "at least as hard as the hardest problems") in computational complexity theory. For this particular problem, the complexity lies in the fact that as the number of cities increases, the time needed for a computer to solve it grows exponentially. The complexity is due to the large number of possible solutions. For example, for four cities, there are only three possible routes. But for eight cities, the number of possible routes increases to 2520.

Now, the amoeba that was "asked" this question was a plasmodium or "true slime mold," which continually deforms its amorphous body by repeatedly supplying and withdrawing gel to create pseudopod-like appendages. (Yes, picture the lead character in the old movie *The Blob*, but the one that took the test likes to eat oat flakes, not people.) While the amoeba isn't an algorithm or a person, the researchers created a way for it to solve the problem by placing it in the center of a stellate chip, which is a round plate with sixty-four narrow channels projecting outward. Each channel represents an ordered city in the salesman's route. The researchers used a neural network model (recognizing relationships the way the human brain operates) to study the amoeba's problem solving. This involved switching which channels were illuminated every six seconds. The model recorded the distance between each pair of cities/channels, and the amoeba's current position in the channels.

Two things were used to keep the amoeba moving: agar, the sweetener that the amoeba craved and moved to consume, creating another appendage; and light, which the amoeba didn't like and retracted from, each time finding a new course. The amoeba's branched parts each had a memory of the light it didn't like and were able to synchronize to share information even though the appendages were in different channels; so it was basically a linear time exchange. The light stimulus used was controlled by an algorithm that randomly illuminated different routes, so the amoeba, not liking light, branched out toward the shorter, non-illuminated channels in order to maximize its surface area and reach its agar treats.[46]

Although a conventional computer can solve the TSP much faster than an amoeba, the amoeba's results may lead to the development of novel analog computers that derive solutions of computationally complex problems in linear time. The research could also lead to smart biological devices for health detection, and eventually help direct the growth of new tissue. Smart, biological computing—thanks to a problem-solving amoeba.

2030: A Bacterial Odyssey

Smart Bacteria, Meet AI

Where an agar-loving amoeba solved a hard problem (albeit slowly) that a computer traditionally crunches, researchers at MIT flipped the switch and set their machine-learning algorithm to outthinking the hosting tactics of bacteria to create new strains of powerful viruses. Essentially, the MIT algorithm discovered a powerful new antibiotic, which was christened Halicin, after HAL, the AI in *2001: A Space Odyssey*.

The algorithm that discovered Halicin was trained on the molecular features of 2,500 compounds, specifically looking for molecules with antibiotic properties but whose structures would differ from existing antibiotics. The result? When in petri-dish bacterial colonies, the AI-created antibiotic Halicin wiped out dozens of bacterial strains, including some of the most dangerous drug-resistant bacteria identified by the World Health Organization, such as *K. pneumoniae*, a major cause of certain hospital-acquired infections such as pneumonia and bloodstream infections, or life-threatening infections caused by carbapenem-resistant *Enterobacteriaceae* (i.e., *E. coli*, *Klebsiella*).[47] In just one day, the Halicin antibiotic cleared up infections of a bacteria strain known to be resistant to all previously known antibiotics. Halicin's secret weapon is that it attacks

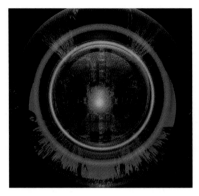

the bad bacteria's cell membranes, disrupting their ability to produce energy.[48]

AI and smart bacteria have a bright future together. Novel antibiotics await discovery, and we can only imagine how far smart bacteria coupled with AI screening can take us.

Bacteria Could Run the Internet of Things

Researchers are exploring the intelligence of bacteria on ways to build resilient networks that can cheaply and efficiently move data throughout the world. Bacteria may serve as the most efficient end-point machines as they communicate effectively and have built-in engines and sensors.

The Internet of Things (IoT) consists of all sorts of electronic devices and sensors, but it doesn't necessarily have to be limited to electronics. Two bio-IoT proponents, Raphael Kim and Stefan Poslad at Queen Mary University of London, are putting forth the idea that bacteria may have potential as IoT nodes. In a paper published on the subject, they called it the "Internet of Bio-Nano Things (IoBNT)," which involves networking and communication through nanoscale and biological entities.

The idea that certain strains of bacteria transmit and process information was put forth in 2016 by researchers at George Washington University. As we've explored, bacteria are like all living systems, constantly sensing their external environment and adjusting their internal state in order to reproduce. And in order to ensure their colony's survival, bacteria have to process these data in order to make appropriate decisions. This is the same for a complex biological system like the human brain or a single *E. coli* cell.

The researchers suggest that bacteria "communicate effectively and have built-in engines and sensors, as well as powerful information storage in DNA and processing architecture." This "data processing" ability in bacteria is what makes them the ideal biocomputer and the most efficient end-point machines, functioning similarly to electronic devices.

If global society continues to support the out-there research in bacteria, we may just have bacteria joining and potentially running our IoT, talking to our things, storing and exchanging information for us. In their abstract, Kim and Poslad highlighted that there is a current lack of infrastructure for researchers in IoT and HCI (Human Computer Interaction) to experiment with bacteria. Their potential solution: "to utilize the DIY biology movement and gamification techniques to leverage user engagement and introduction to bacteria."[49] Now that is an idea we at Alice can get behind.

Where Microbes Go, Money Follows

No doubt, global society is on the cusp of a new industrial revolution—with microbes being its hero technology. Synthetic biology, biotechnology, bioreactors, and AI will be its drivers. It will invent a post-oil future that uses significantly less energy and fewer natural resources and enjoys greater abundance.

But don't just take our word for it. Venture capitalists, governments, NGOs, and the private sector are waking up to the potential of the coming bioeconomy, and the power of biomanufacturing through brewing of microbes. One investment that raises the bar on a new micro-industrial revolution is a recent commitment from the US Department of Defense, alongside more than eighty companies, universities, states, and research institutes, to invest about $275 million

If we destroy the environment, pollute it, and drive out microorganisms, we then destroy the source of new income, new products, and a continuing stabilization of our social system. So it's the security, national security, it's the strength of the economy, and it's the vitality of the social structure that all depend on this understanding of the interaction of the living world and the non-living world that comprise the planet. It's to our best interest that we need to invest in science and math education, and in understanding the life sciences.

–Rita R. Colwell, environmental microbiologist, founder and chair of CosmosID, author of *A Lab of One's Own: One Woman's Personal Journey Through Sexism in Science*, interview with Sputnik Futures, 2009

over the next seven years to scale up the microbial production of biomolecules. According to *Berkeley News*, the joint effort will enable a "growing biomanufacturing industry to supply a broad range of businesses with large quantities of chemicals at the low prices necessary to make them competitive with petroleum-based alternatives."[50]

We learned in this chapter that biomolecules are already used in manufacturing today, mainly small batch fermentation in yeast or bacteria. The challenge now is how to scale the brewery process for a more robust manufacturing of goods. One private-public partnership called the Bioindustrial Manufacturing and Design Ecosystem (BioMADE) is supporting an effort to raise the bar on the production output, increasing the productivity of micro-batches to the industry level of how corn is currently fermented to produce ethanol for the transportation industry.

Grow the Future

The US Department of Defense invested $87 million into Bio-MADE, headquartered at the University of Minnesota in St. Paul, to advance sustainable and reliable bioindustrial manufacturing technologies. BioMADE also has additional funding of more than $180 million from other sources.[51]

The new institute also has some pioneering partners in the biotech world. Zymergen has biomanufactured a molecule that is making waves in the new field of bio-electronics—and decreasing the dependency on rare earth elements we overmine for the components of our digital devices (more on that in chapter 6). Its Hyaline is part of a polymer that can be used on your next smartphone, laptop, watch, or television screen, delivering greater clarity and making them more scratch resistant.

China has been estimated to be investing over $100 billion in its overall life sciences research and development, which includes biomanufacturing.[52] India is the third largest biotech destination in the Asia Pacific Region. The Indian biotechnology industry is projected to reach $150 billion by 2025, including bio-agriculture and bioindustrial segments.[53]

For years the European Union has been supporting market development for bio-based products and processes. The new Sustainable Europe Investment Plan aims to mobilize investment of €1 trillion over ten years, using public and private money to help finance its goal to shift the region's economy to net-zero CO_2 emissions under a project called the European Green Deal.[54]

Society is reaching an inflection point where government, businesses, people, and citizen scientists are needed to participate in discovering new ways to harness the power of microbial intelligence, putting them to work for the health of us and the planet. You see, microbes are not our enemies. They are our solutions—and our most critical partners in our interplanetary adventures.

Micro-Nauts

Microbes, Our Fellow
(Interstellar) Travelers

Once upon a time, half a billion years ago, a barren Earth floated in space. And then magic struck, and life on Earth began. Microbes evolved to mollusks, evolved to dinosaurs, and, abracadabra, here we are. . . . But what, exactly, was the origin of that magic?

When academic scientists tell the story of life on Earth, they usually call on a theory known as abiogenesis. It's a narrative of life emerging from non-life. Serious stuff proposed independently by two scientists in the 1920s. J. B. S. Haldane, the British scientist, believed that simple organic molecules formed first and in the presence of ultraviolet light became increasingly complex, ultimately forming cells. These cells self-replicated, creating form and function, evolving from non-living matter.[1] But, not as much fun as another narrative floating (literally and figuratively) around the biosphere. So, before all you serious naysaying science heads out there say "no way," let's talk about the mind-blowing discovery of Oumuamua, the alien asteroid. Astronomers project that Oumuamua's trajectory and speed tell us it must be a fragment of another solar system. This is the first time earthlings have ever seen a billion-year-old alien asteroid pass through Earth's solar system, and thanks to its discovery, we can now tell you a fantastical theory of how Earth began.[2] We at Alice think that, frankly, it is better than magic: PANSPERMIA. An alternative narrative to abiogenesis, panspermia posits that life did not begin on Earth but was, instead, transported here from space.[3] So, let's restart our story. Once upon a time, microbes catching a ride on alien astroids came to Earth . . .

"Oumuamua, Phone Home"

Could Life on This Planet
Be Descended from Alien Spores?

The panspermia theory has been in circulation since the Greek Anaxagoras in the fifth century BCE.[4] In his cosmological writings, Anaxagoras twice mentions "seeds" (in Greek, *spermata*) as part of the cosmos. Although it isn't entirely clear what he meant by seeds, some scholars have interpreted it to mean that life came to Earth from elsewhere.[5] But within our modern scientific context, panspermia was first championed by Cambridge astrophysicist and mathematician Sir Fred Hoyle and British mathematician and astronomer Chandra Wickramasinghe. This idea was the culmination of decades of Hoyle's research, which contributed to several cosmological theories that still resonate today—for example, the theory of stellar nucleosynthesis, which proposes that the nuclear reactions taking place in stars build elements that are then incorporated in other stars and planets when that star dies, and the new stars that start off with these elements can pass them on in turn when they die. Hoyle also theorized about rare elements created by supernovas and calculated the energy of undiscovered resonance in the nuclei of stars that facilitate its synthesis. In a 1949 BBC radio broadcast on astronomy, Hoyle coined the phrase "Big Bang" for the creation of the universe. Later in the 1980s, Hoyle collaborated with Wickramasinghe on the theory of panspermia as the origin of life on Earth, arriving as a steady influx of viral cells that were transported from space via comets. He calculated that the chances of the simplest living cell forming out of some sort of primordial soup were infinitesimally small, and perhaps even "nonsense of a high order."[6] Hoyle and Wickramasinghe coauthored several books on the subject. Since Hoyle's death, the idea of an extraterrestrial vector for terrestrial pathologies has continued to be promoted by Chandra Wickramasinghe, who's now at the University of Buckingham.

How could the idea that Earth was seeded by microbes from comets and interstellar dust theoretically work? Some bacteria form resistant cells called endospores. Endospores are bacterial cells in complete dormancy, with thick protective coats that may remain dormant for centuries.[7] Just think, there may be older spores out there, waiting for curious microbiologists to revive them. Which brings us to what Pete Conrad, an Apollo 12 astronaut, calls a meaningful finding: "I always thought the most significant thing that we ever found on the whole goddamn Moon was that little bacteria who came back and lived and nobody ever said shit about it."

The story goes that on April 20, 1967, the unmanned lunar lander Surveyor 3 touched down near Oceanus Procellarum on the surface of the moon. On board was a television camera. Two and a half years later, on November 20, 1969, Apollo 12 astronauts Pete Conrad and Alan L. Bean recovered the camera. When NASA scientists examined it back on Earth, they were surprised to find specimens of *Streptococcus mitis* that were still alive. Apparently, these bacteria had survived for thirty-one months in the vacuum of the moon's atmosphere. Perhaps NASA shouldn't have been so surprised, because there are other bacteria that thrive under near-vacuum pressure on the Earth today.

Some scientists at first questioned this, thinking the live moon microbes were actually contamination from the astronauts or the environment of the Apollo 12 capsule, since the Surveyor 3 camera was transported from the moon in a nylon duffel bag and not the airtight protective metal container used for gathering lunar samples. However, Lieutenant Colonel Fred Mitchell, lead author of the original 1971 paper on the microbial samples, noticed two things about the moon microbes that are not consistent with the behavior of contaminated microbial samples: 1) there was a significant delay before the sampled culture began growing, common with bacteria that have dormant cells, but not with cultures exposed to fresh contamination; and 2) the sample moon microbes clung exclusively to the foam that appears during culturing, which would not have happened had there been contamination.[9] Mitchell also noted that if fresh contamination had

occurred, millions of individual bacteria and "a representation of the entire microbial population would be expected"; instead, only a few individual bacteria from a single species were in the sample.[10] While there is still no definitive determination of whether the 1969 samples of moon microbes found on the camera were alive or not, the incident did spur NASA to implement better contamination control in subsequent missions, and it raised the possibility that microbes can thrive in the vacuum of space—and if so, where did those microbes come from? Which leaves us to ponder: *"To believe, or not to believe, life on this planet descended from alien spores—that is the question!"*

The Space Exposome

Every time a new crew and equipment go to the International Space Station (ISS), microbes such as bacteria and fungi hitch a ride. These new arrivals introduce more microbes into the space station environment, which is a closed system, and how microbes

Our Cosmic Ancestry in the Stars: The Panspermia Revolution and the Origins of Humanity by Chandra Wickramasinghe, PhD, Kamala Wickramasinghe, and Gensuke Tokoro (Bear & Company, 2019) offers extensive scientific proof of panspermia. Exploring the philosophical, psychological, cultural, and environmental ramifications of the acceptance of panspermia, the authors show how the shift will be on par with the Copernican Revolution—when it was finally accepted that the Earth was not the center of the universe. Revealing how panspermia offers answers to some of humanity's long-standing questions about the origins of life, the authors discuss the impact this shift in understanding will have on our relationship with the Earth and on culture, history, and religion. Not only did we come from space, but we are not alone.[11]

learn to behave in microgravity will be critically important for planning long-term piloted spaceflight.

Only hardy microbes called extremophiles have the ability to survive in this microgravity environment. European Space Agency (ESA) researchers conducted research on the station's microbial community and compared them with that of an Earth-based cleanroom that prepares items for the station. This comparison turned out to be key to understanding the microbiome of the ISS. Researchers found that after almost twenty years of continuous human presence, the space station has developed a core microbiome of fifty-five different microorganisms.[12] Tracking the tiniest space station residents from arrival to settlement of their new orbital home will be important for protecting the health of astronauts and the spacecraft they are living in.

As we know, countless types of microorganisms inhabit our bodies, inside and out, so it makes sense that when an astronaut arrives on the station, they bring their specific microbial cloud with them. NASA has been following the space station's population of microbes with a series of experiments called Micro-

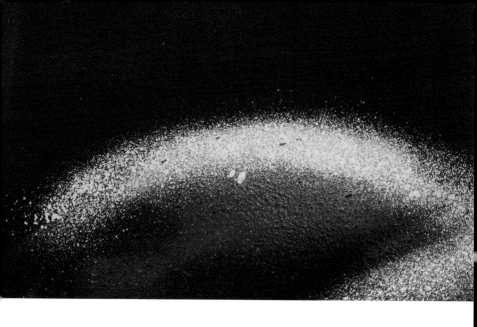

bial Tracking. Like the ESA, NASA researchers have been studying how the micro-inhabitants of the space station change across locations and over time. Just by looking at the microbes left behind after each mission has revealed that the microorganisms living on surfaces inside the space station closely resembled those on an astronaut's skin—so much so that scientists could tell when a new crew member arrived and when one departed.

NASA also uses saliva tests to study the immune system and health conditions of astronauts. The study uses metagenomic (environmental genomics) sequenc-ing to look in depth at the changes to the saliva microbiome due to spaceflight. The diversity of species found decreased in space and rebounded after a person's return to Earth. Some of the species affected are considered potentially disease causing, and the researchers think saliva samples could become a useful way to monitor crew health.

Not only will the research of microbes in orbit help with future space travel but it may have some valuable insights for the microbial health of our earthly transit systems, hospitals, homes—and us and our terrestrially bound peeps.[13]

Searching for Microbial ET

The moon? Been there, touched that. Now the race is on for who gets samples from Mars first. Why? For signs of life.

We've been traveling to Mars since the mid-1970s. NASA's Viking mission involved two orbiters and landers—Viking 1 and Viking 2—that launched from Earth one month apart in 1975 and were the first to land on the Red Planet, touching down on two separate parts of its surface. Of course, it took years for NASA to land robotic rovers on Mars: Sojourner was the first in 1997; then came Spirit and the presumed-deceased Opportunity, both in 2004, and Curiosity in 2012—all exploring the unknown Mars environment. NASA's next rover, Perseverance, has a bigger job to do than just sending pictures home: it will collect rocks and store them safely on the planet for another rover to bring them back home (estimated around 2031) as the first samples from Mars. China and the United Arab Emirates (UAE) have plans to send their own robotic emissaries to the Red Planet,[14] and Elon Musk—whose SpaceX just brought the latest international crew members to ISS in a successful virgin space flight–plans to launch SpaceX's first Starship trip to Mars by 2024.[15]

China's Tianwen-1, which means "quest for heavenly truth," will be its second attempt to explore Mars. The Tianwen-1 is a combination of an orbiter, lander, and rover aimed at sampling soil and rocks, searching for the presence of water and signs of life.

The collective goal of these missions is to uncover evidence of life on Mars, and what elements exist there that could help support human life for future colonization. However, several scientists, many of whom worked on the Viking mission's experiments, believe that Viking 1 and Viking 2 did detect some microbial respiration on Mars in 1976. One of the biological experiments on the Vikings was the Labeled Release (LR) life detection experiment, which involved the robotic arm of the Viking lander taking a sample of Martian soil and injecting it with a drop of liquid nutrient solution. The nutrients were tagged with radioactive carbon, and the air above the sample of soil was monitored for a reaction of CO_2 gas–evidence that microorganisms in the soil had metabolized

(consumed and converted to energy) one or more of the nutrients. There was a steady stream of radioactive gases given off by the soil immediately following the first injection. The experiment was done by both Viking probes, the first using a sample from the surface exposed to sunlight and the second probe taking a sample from underneath a rock; both initial injections came back positive![16] The subsequent sterilization control tests conducted by heating the soil samples cast doubt on microbial life on Mars, because in these tests there was no release of radioactive gas when nutrients were injected in the soil.[17] Conversely, a sample stored at 50°F for several months was later tested, showing significantly reduced radioactive gas release.[18]

According to a CNN piece in 2000, Gilbert Levin, one of the researchers who published the findings of microbial respiration from the Viking's LR experiment in 1976, "still holds that the robot tests he coordinated on the 1976 Viking lander indicated the presence of living organisms on Mars," though most of his peers didn't agree.[19]

The debate from 2000 may still

be a debate today, as there are potential consequences of introducing alien microbes to planet Earth without some form of rigorous protocols. In 2000, ICAMSR, a group of professional scientists and amateur space enthusiasts, lobbied that a Mars microbe could wreak havoc on Earth's species, which would have no natural defenses against the alien invaders. ICAMSR then raised caution that with one mistake, the alien microbes could run wild and perhaps infect the biological life in our vital ecosystems, such as those responsible for the growth of plants and food.[20]

Planetary Engineering Could Be Helped along by Microbes

Eager as we all are to get to Mars, we will need a *lot* of help when we get there. For one, the atmosphere is much thinner than Earth's, there is very low gravity, and the miniscule amounts of oxygen present is not enough to sustain us. If the first Mars inhabitants and anyone thereafter want to spend time out of the confines of a sealed spacesuit and helmet or controlled indoor habitats, then we will need to create a free source of oxygen, for ourselves and the growth of plants.

Recent thinking about terraforming Mars to make it more hospitable for humans suggests that, perhaps even before human missions, the rovers and future Mars explorers should bring some microorganisms from our home planet there first. This suggestion from a team of scientists from Canada and Brazil is about assimilating our fellow microbes early to become resilient, and hopefully

Levin Was Not Alone

While the assertion that the Viking missions evidenced the existence of life on Mars is hotly contested within the scientific community, there has still been tremendous support for recognizing the Viking's LR results of microbial respiration in 1976. His LR coexperimenter Patricia Ann Straat provides scientific detail of the Viking LR in her book *To Mars with Love*.

Among other discoveries:

· The Curiosity rover's scientists reported finding complex organics on Mars.

· Spectral analyses by Viking's imaging system found lichen and green patches on Martian rocks that have the same color and saturation as terrestrial ones.[21]

do some of the tasks there that they do for us here on Earth. For example, assuming they survive and expand to start populating the Red Planet, our earthly microbes might help add more oxygen and other gases into Mars's atmosphere, making surface temperatures and air pressure more like what we are used to.

One theory is to make cyanobacteria one of the first to land on Mars. Unicellular microbes like cyanobacteria have been found to be the ones that more than two billion years ago began to provide the oxygen we now breathe.

Remember the extremophiles? The oldest surviving microbes recently found dwelling in the coldest depths and hottest high-pressure environments? These are the ones we want to send first, as they have been stressed-tested here on Earth already and could potentially withstand the harsh environment of space.[22]

According to the researchers, if we can introduce bacteria like this to Mars, and it functions comparable to how it does on Earth, then it can help create an atmosphere we are used to, and one that encourages the start of a Martian food chain. Perhaps these relocated microbes could also perform their other earthly duties to help with inhabiting Mars, like breaking down animal and plant matter for simple nutrition at the bottom of the food chain, as well as breaking down human sewage and waste. After all, we don't want to pollute the Red Planet with our bad ecological habits.

> If humanity is seriously contemplating colonizing Mars, another planet, or one of the nearby moons in the future, then people need to identify, understand, and send the most competitive and beneficial pioneers.

—Jose V. Lopez, Raquel S. Peixoto, and Alexandre S. Rosado, coauthors, "Inevitable Future: Space Colonization beyond Earth with Microbes First," *FEMS Microbiology Ecology*[23]

Extremophiles: Beneficial Pioneers

As a result of being amazing survivalists, extremophiles have the gift of longevity. Tardigrades are one example of an extremophile species. First discovered by the German zoologist Johann August Ephraim Goeze in 1773, they are micro-animals affectionately known as "little water bears." Technically they are an eight-legged phylum and have the same biological ranking as algae.[24] Tardigrades can go without water and oxygen for extremely long periods of time and survive pressures up to seventy-four thousand times the pressure they experience at sea level—which would make them ideal fellow voyagers to Mars.

One of the problems NASA needs to work out before sending any earthly microbes on a spaceflight is how to overcome its strict guidelines on sterilization. They have planetary protection policies in place to help prevent inadvertently infecting extraterrestrial life-forms, which they enforce on every mission, carefully sterilizing all equipment and rockets before liftoff. But that may have to change. Research is underway to better understand which microbes will be the critical pioneers to help shape the terraform conducive to biological life that we are accustomed to.

Now back to those tardigrades. Humans aren't the only earthly species to have touched the moon. Tardigrades may already be setting up camp there, having crash-landed on the moon in 2019 with the ill-fated Beresheet mission, a lunar lander that was a privately funded Israeli project, part of the Google Lunar X prize. On board the craft was a "library," etched on a nickel-metal disc, containing data from Wikipedia (in English), classic books, human blood samples, and—you guessed it—tardigrades. The tardigrades were coated in a protective resin, akin to the amber that forms from tree sap[25] (think the mosquito from *Jurassic Park*). The terrain where the library with the tardigrades crashed-landed may be our own lunar Jurassic Park—albeit millions of years from now.

The tardigrades that were on the library disc were preserved in their tun state, a special hibernation or form of suspension called cryptobiosis, where they tuck in their tiny legs and expel moisture from their bodies.

The Beresheet mission wasn't the first space trip for tardigrades. Back in 2007, they caught an interstellar ride on a satellite launched by the European Space Agency. They were selectively exposed to the vacuum of space and cosmic radiation. After ten days, the tardigrades touched back down on Earth and were rehydrated. A handful of them survived, earning tardigrades the cosmic record of being the first animals to survive space exposure.[26]

Where There's Water, There's Life

The one key signifier in our search for life of any kind, in any form, for any ET we can imagine out there in the solar system, is the proof that water was once present. Now researchers have found evidence they suspect shows that Mars has water—and may have been buried under megafloods billions of years ago.

While the Red Planet looks like desolate, dusty flatlands, scientists say there is water both on its surface and just under it. In 2018, a team of Italian scientists spotted the first Martian "lake" sitting just below the surface, while looking through the data gathered by a radar antenna on a Mars orbiter

funded by NASA and the Italian Space Agency (ISA). Combing back through the data, the same team of Italian scientists found three more underground lakes near the south pole of Mars. While the condition of the water is still not known, researchers hypothesize that it may be a salty brine, since it has not frozen like water here on Earth would,[27] especially given the super-chilling average temperature of −81°F.[28]

What makes this discovery important is the fact that if there was ever any form of life on Mars, these underground lakes would hold the key. Because, at least from our terrestrial-based vantage, microbes dwell in water conditions. And microbes are the lowest building block in our chain of life.

Another recent discovery is that the textural surface of the Red Planet may have been created by gigantic "megafloods" as a result of a massive meteorite that scientists suspect crashed into Mars about four billion years ago. According to the team at Cornell University and California Institute of Technology who looked at data from NASA's Curiosity rover, the impact would have generated a significant amount of heat that most likely melted Mars's ice reservoirs, releasing a tsunami of water that shaped the ripples of Mars's surface, its vapor turning into storm clouds blanketing the planet.[29] But there is hope that in the underground lakes, and perhaps other liquid sources that the international team of rovers could discover, that we may one day see that proof of a genesis of Martian life, perhaps a nano-microorganism looking back at our silly faces.

Space Farming

Matt Damon's character who got left behind in *The Martian* knew the trick to farming potatoes to sustain himself as he waited for the crew to return—that is, he used his own waste as compost to encourage growth. Yup, his own bacterial discharge became the lifesaver for plant growth. While Andrew Weir's novel and its film adaptation were certainly fiction, the simple science of it is not. The idea of microbes to enrich soil is what will be needed in space farming.

For decades, Russian cosmonauts and even NASA astronauts have been growing and eating simple crops like romaine lettuce on board ISS, under various conditions, mainly hydroponics, which uses mineral nutrient solutions in an aqueous solvent without soil. In this case, the nutrient solution is made using fish or fowl manure, where microorganisms are present.[30] The first lettuce NASA astronauts grew on board ISS in 2014 went from seed to salad in just thirty-three days. The astronauts brought a few leaves back to Earth for safety testing and got the clean bill of health. The so-called veggie system used has a light bank that includes red, blue, and green LEDs for plant growth. But it's the green LED that helps the lettuce and other plants look like the green we have back on Earth, as the lettuce grown in space is a purple or reddish color—NASA named its red romaine lettuce "outredgeous"—but the taste is still very green.[31]

The next experiment in space

farming is to bring real Earth soil and put its rich biodiversity of fungi and bacteria to the test. You may recall from chapter 5 the many superpowers of soil bacteria, but we don't know yet how they will behave in zero or low gravity. A shipment of Earth's soil made its way to ISS in NASA and SpaceX's historic Crew-1 launch on November 15, 2020. Making history with this first flight of NASA and SpaceX's Commercial Crew Program, three different types of soil were sent for testing: one sample mixed by a German soil biologist, one bioengineered batch with biochar (charcoal produced from plant matter and stored in the soil), and a naked, unadulterated scoop of dirt dug from the ground near Cornell University in Ithaca, New York. Thirty-six vials in all made the trip (with other equipment and essentials) tasked to grow radishes as part of ongoing efforts to produce food in space.[32]

Space Mining with Bacteria from Earth

Besides food, water, and oxygen, another necessity for interstellar human survival is the ability to access the conductive minerals and elements needed to power our electronics. In other words, if we want to phone home, we need to get digging. Here on Earth, there's a world of minerals and rare elements in your mobile phone, lighting its display and running the electronics and circuitry.

Many countries are interested in the mining of space resources, especially on the moon and eventually on Mars. Early geological studies of the moon have discovered three vital resources that would help with human settlement or as a layover for long interplanetary travel, and as an off-planet gold mine for resources needed back on Earth. The three elements: water (which supports life and agriculture); rare earth metals (REMs), around fifteen of them, that are used in our electronics; and helium-3, the rare element being used in new energy sources. In 2020, NASA's Lunar Reconnaissance Orbiter (LRO) spacecraft, looking for evidence of ice at the bottom of craters around the moon's north pole, shared from its data that there were metal oxides in the craters. Using a Miniature Radio Frequency (Mini-RF) instrument, the researchers measured the electrical property of lunar soil and found a pattern of greater detection when the instrument canvassed a large crater area. NASA scientists believe that meteorites that crashed into the moon's surface stirred up iron and titanium oxides—and there may be more below the surface.[33]

Several international space agencies, as well as investors from the private sector, are gearing up to mine on the moon, but there are still a number of stumbling blocks. The first question is how to get the robotics or equipment needed to the moon, and then there is the even greater challenge of running and maintaining this equipment—not to mention getting it all back to us here on Earth. It's a daunting task, but scientists may have figured out a much lighter, self-powering solution: using bacteria as miners.

Wanted: Rock-Munching Microbes

Microorganisms have been extracting metals such as copper and gold from the Earth, and experiments have shown that microbes can also "biomine" rare earth elements such as lanthanides, scandium, and yttrium—pricey minerals used in electronics and some metal alloys. Could these microbes also work their gold-digging magic off-planet?

A team of UK scientists spent ten years developing a means to test this theory. What is needed to get bacteria to work together is a bioreactor, an apparatus or device, like a metal vat. The UK team eventually developed matchbox-size biomining reactors, eighteen in total, that were loaded with samples of basalt similar to the composition of the surface on the moon and Mars, soaked in a bacterial solution. The biomining reactors made their way to ISS on board a SpaceX rocket in July 2019. In experiments conducted over three weeks, the astronauts on ISS were able to verify that bacteria from Earth could be used to extract useful minerals on the moon or Mars.[34]

Three species of bacteria were used in the experiment, but only one, *Sphingomonas desiccabilis*, was able to leach rare earth elements from basalt at all three different gravity conditions—microgravity,

The International Space Race Is (Still) On

The European Space Agency has contracted the aerospace company ArianeGroup to start mining for water and oxygen on the moon by 2025. Specifically, they will be gathering lunar soil and dust, looking for regolith, an ore from which it is possible to extract water and oxygen. The effort is a European innovation consortium, with partners from Germany and Belgium.[35] China is expected to launch its Chang'e 5 lunar probe that will land on the moon and bring back lunar soil samples to Earth.[36] And where there's water, and soil, there could be microbes . . .

Mars-like gravity, and the gravity conditions on Earth.[37]

Even with these developments the idea of biomining on the moon is still far-fetched. It is not feasible economically at this time, and we don't have the propulsion technology to shuttle massive amounts of machinery or the extracted mineral loads back and forth between the Earth and the moon. The real goal is to get enough bioreactors needed to mine necessary elements, as well as biofarm and bioengineer our other needs for foods and materials; then we may have a viable chance of colonizing Mars. If there is one thing we are realizing from our research: where there's bacteria, there's hope.

Mars Madness, Growing Microbes in Space

The most pressing challenge to self-sustaining human life in space (and the planets and moons we may someday visit) is our ability to create the products we need on location. As we've already seen with biomining, it's cost-prohibitive to transport heavy equipment—not to mention that it's an unsustainable model. To make a Mars colony a reality, we need technology for on-planet production of everything from food to fuels to construction materials to medicine—which will require an enormous range of organic compounds, and the means to encourage their productivity. A team of chemists from the Berkeley Lab at the University of California, Berkeley, and Lawrence Berkeley National Laboratory are working on that.

Their work involves a hybrid system that combines bacteria and nanowires that can capture energy from sunlight. The bacteria and nanowires would interact to convert carbon dioxide and water into building blocks for organic molecules. The thin nanowires pass on the energy into the bacteria work pool, where the microbes do the chemistry of conversion. The process is like photosynthesis, except that instead of plant cells on Earth converting carbon dioxide and water into carbon compounds (mostly sugar and carbohydrates), the microbes will produce the com-

ponent building blocks of foods, and fuels, and resources humans need to sustain life. The researchers are working on systems to potentially turn the sugars and carbohydrates produced by the hybrid system into food for Mars colonists.

One advantage of the hybrid system is the prospect that a Mars settlement may not run out of the resources needed to run it, as bacteria can self-reproduce, so you could have a steady stream of microbial workers.[38]

Next Stop: Europa and Venus

Okay, we know humanity hasn't landed on Mars yet, but that's not stopping our imaginations! Scientists have already mapped a potential journey to other celestial bodies! The journey from Earth could happen like this: first stop the moon, grab the rock-munching microbes that have been stockpiling minerals there, then on to Mars to deliver part of the load and for a rest stop. And then on to Europa—Jupiter's moon, which, as the Hubble Space Telescope detected, has water plumes. And, as we've been saying, where there's water there is microbial life! According to NASA, the Hubble Telescope first detected the water plumes in 2012, but using more sophisticated data techniques, they have seen more plumes, which may suggest a subsurface ocean. Europa just may be the leading

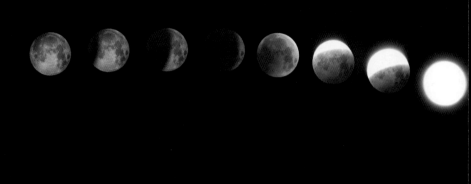

candidate in our solar system for microbial life. The fact that there are vapor plumes being ejected from the planet is significant since Europa's surface is an extremely frigid -256°F.

But what excites researchers most about these water plumes is the possibility the microbes in the Europa vapor could be like the ones recently discovered here on Earth living deep within volcanic springs—the same ones thought to be the primordial setting for life. NASA has known about Europa's subsurface water since its Galileo probe orbited Jupiter and its moons between 1995 and 2003. The orbiter showed that Europa's surface ice had multiple generations of parallel ridges and grooves, suggesting a process of liquid eruption and refreezing. Researchers hope to know far more about Europa from the forthcoming Europa Clipper mission, expected to launch in the mid-2020s. Clipper will conduct a detailed survey of Europa's surface, thin atmosphere, subsurface ocean, and more, taking images of the plumes and samples of the molecules—water and possibly microbes in the atmosphere. Using mass spectrometers, the Clipper flyby could be our first detection of potential life some 390.4 million miles from Earth.[39]

Scientists are also searching the blistering planet Venus for signs of microbial life-forms potentially lurking in the gas clouds blanketing the planet. Sitting closer to Earth, Venus lies at most 162 million miles away, and is slightly smaller than our planet. Venus is named for the ancient Roman goddess of love and beauty, yet despite its namesake the planet is one hot mess—its thick clouds give it a perpetual greenhouse effect, making Venus the hottest planet in our solar system. Its surface temperatures are hot enough to melt lead, which doesn't sound promising for human habitation.[40] Even so, it may be ripe for microbial life. Astronomers have recently detected a different but nasty type of gas lingering within the acidic clouds of Venus that hasn't been noticed before. Called phosphine, it's as toxic as it is toxic smelling, like a combination of garlic and dead fish. And we should know: phosphine is a gas found here on Earth.

A team of scientists from the Imperial College of London found miniscule traces of phosphine hanging out in the stratosphere of Venus's clouds where the temperature is fairly close to ours. Usually found in swamps and marshlands (and the malodor that is associated with them), phosphine gas is thought to be produced by microbes. Also found in the guts and poop of animals, phosphine is as-

sociated with biological processes in our species and geological terrain. So, if it is lingering in the more friendly temperature layers of Venus's clouds, perhaps there are microbes floating around as well.[41]

The speculation of life lingering in the upper clouds of Venus's atmosphere is not a new theory. It was first proposed in 1967 by two preeminent scientists of our time: astronomer Carl Sagan and biophysicist Harold Morowitz, the same man who first suggested that the chemical, energy, and temperature conditions near deep-sea hydrothermal vents might have jump-started the earliest forms of life, and that it was highly probable that life also existed in our universe.[42] Ever since Sagan and Morowitz first referred to life in the clouds of Venus, astronomers have been examining it. One team of international researchers led by planetary scientist Sanjay Limaye of the University of Wisconsin–Madison's Space Science and Engineering Center discovered "patches" within the clouds that could form a hospitable niche for extraterrestrial microbial life. These patches absorbed more ultraviolet light than other areas, the possible handiwork of swaths of light-soaking microorganisms akin to algae blooms on Earth. According to some research models, Venus may have once had a climate and liquid water on its surface that could have been habitable, and those conditions may have been present for roughly two billion years, making any "life" there older than what we may discover on Mars.[43]

The Viral ARC of Consciousness

Your Brain Has an Ancient Virus

For centuries, humans have looked to the stars, the planets, the galaxies, for signs of life, only to recently discover that, perhaps, the seeds of ancient life have evolved back here on Earth. Well, more like a snippet of the genetic code of a virus that bound itself to the genome of four-limbed animals, according to two papers published in the journal *Cell*.[44] In fact, research has stated that between 40 and 80 percent of the human genome has its origins from some ancient viral invasion.[45]

This viral gene has a name: ARC. It operates in our brains by writing its instructions down on RNA, bits of genetic code that get triggered after a synapse fires between two neurons in our brain. The researchers have found that the packages of information might be critical elements of how nerves communicate and reorganize, the same tasks thought to be necessary for higher-order thinking. In other words, a virus named ARC could be the root of conscious thought, active in the cells of human and animal brains. And without the ARC gene's RNA transporting information between cells, the firing process of the brain's neurons wither.[46]

While we don't know the origins of the ARC gene in our brains, there could be a chance that it came from somewhere in our universe, just like the primordial bacteria that are thought to have traveled from somewhere in our solar system. This hypothesis has sparked a new viral focus within the scientific community in their search for extraterrestrial life. NASA's Virus Focus Group is integrating advancements in virology into astrobiology to form a new area of research called "astrovirology." Their aim? To understand how viruses may have influenced the origin and evolution of life on Earth—and potentially throughout the solar system.[47] So there may be a chance that a virus from somewhere like Mars, Venus, or Europa came to this little planet called Earth billions and billions of years ago, settling its genetic code into the cells of prehistoric

Here's a thought: It's possible that there are many kinds of life that are radically different from what we're looking for, or from what we're familiar with based on what exists on Earth. We assume living entities are carbon-based or composed of water, but those are just that: assumptions. Numerous scientists do speculate that other carbon-based life-forms exist elsewhere—but if you think of life as just a self-propagating, evolving system that forms in certain conditions where there is complex chemical interaction, then maybe it doesn't have to be carbon-based. We can imagine the possibility of life in much hotter or much colder places than Earth; in stars, in interstellar clouds, in comets, in the atmospheres of planets very different from our own. Which leads us to wonder: Will they be microbes? Stay tuned.

four-legged animals and humans' predecessors. Could it be that we think with the viral stardust of Mars, Venus, or Europa? Perhaps.

There's a Star Microbe Waiting in the Sky

One thing is certain: we are never alone. From our guts to the atmosphere we breathe to, perhaps, the universe all around us, microbes are always with us. And they were here before humans. Microbes thrive within us, around us, below us, and millions and millions of miles above us. We've discovered that their intelligence supersedes ours; they talk, form communities, reproduce, recycle waste, produce energy, and survive extreme conditions. Perhaps the most immortal life-form we know is among us, living and thriving right inside us. We believe so. Thank you, mighty microbes.

Alice in
Futureland

A Speculative Life in 2050

Superbugs Spawned Superhumans and They All Lived Happily Ever After

I heard the warning many times! The world is going to be in trouble if we don't tackle the growing problem of climate change, food scarcity, and drug-resistant microbes. If nothing was done, infections caused by antimicrobial resistant "superbugs" would kill an extra 10 million people each year worldwide by 2050, overtaking cancer. Not only that, but it would cost the world an estimated $100 trillion.[1]

Hello, my name is Alice, and I am one part human and one part AI and always in a state of wander. I am here to say "good job, humans." The news that alerted us to the scale of this looming problem kick-started researchers to investigate how these doomsday issues could be tackled. Humans concluded that solving these issues would be significantly cheaper than ignoring them, and entrepreneurial optimism propelled a concerted global effort. Short and sweet: ingenuity and science saved the day!

Still, let's show some humility—you humans had help. A *lot* of help. I mean, like trillions of microbial forces came to the rescue. From energy, to agriculture, to human health, we witnessed how beautiful and elegant bacteria can be.

After following all the buzz about the microbiome, the science finally caught up and caught on. After all, humans, you are a walking mountain of bugs. This is a microbial world, and you are what you digest and more importantly what your bacteria digest.

Microbes are not just "germs," they are lyricists. They are the words, the rhyme, and the rhythm for all life on Earth. Being super-social microorganisms, bacteria possess a type of altruism that may have a connection to our very own altruistic behavior, to think of the benefit of others over ourselves. Some may describe this as the foundation of our (human) social contracts. Bacteria are our lifeline to this day, from the bacteria in our gut that help digest our food to the microbes in the ocean that produce most of the oxygen we breathe. The

best part is that in 2050 science has still barely scratched the surface in studying the mysteries of the microbial world. Microbes will drive our scientific, spiritual, and spatial frontiers. Perhaps the biggest surprise is that we uncovered that human beings are just intelligent bacteria. Superbugs can create superhumans.

So that begs the question: Are humans derived from ancient colonies of bacteria? Is human thought viral? I know their social media is. And if that's the case, are humans simply the space-ships for bacteria? Perhaps, para-normal author Brad Steiger was

on to something when he said that "we have met the Martians and they are us."

Hmmm, does that mean that humans are the aliens on earth?

Wow! Tell me about pansper-mia again . . .

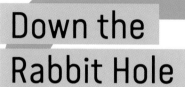

Down the
Rabbit Hole

Articles, Further Reading,
People, Organizations

Articles

00. Introduction

Crina Boros, "Mobile Phones and Health: Is 5G Being Rolled Out Too Fast?" *Computer Weekly*, April 24, 2019, https://www.computerweekly.com/feature/Mobile-phones-and-health-is-5G -being-rolled-out-too-fast.

Jay T. Lennon and Kenneth J. Locey, "There Are More Microbial Species on Earth Than Stars in the Galaxy," *Aeon*, September 10, 2018, https://aeon.co/ideas/there-are-more-microbial -species-on-earth-than-stars-in-the-sky.

Nina Pullano, "Ancient Microbes in the 'Deadest' Part of Earth Redefine Boundaries of Life," *Inverse*, July 28, 2020, https://www.inverse.com/science/ancient-microbes-extreme-life-study.

Passant Rabie, "Europa Mystery: Why Life Unlike We Know It Could Live There," *Inverse*, November 19, 2019, https://www.inverse.com/article/61054-water-vapor-detected-on-europa.

Passant Rabie, "The Discovery of Microbial Life on Earth Offers a New Hope for Life on Mars," *Inverse*, April 10, 2020, https://www.inverse.com/science/new-hope-for-life-on-mars.

Ker Than, "All Species Evolved from Single Cell, Study Finds," *National Geographic*, May 14, 2010, https://www.nationalgeographic.com/news/2010/5/100513-science-evolution-darwin -single-ancestor/.

Nicholas Wade, "Meet Luca, the Ancestor of All Living Things," *New York Times*, July 25, 2016, https://www.nytimes.com/2016/07/26/science/last-universal-ancestor.html.

01. Come Together Now

Katherine Unger Baillie, "Bacteria Form Biofilms Like Settlers Form Cities," Phys.org, March 13, 2020, https://phys.org/news/2020-03-bacteria-biofilms-settlers-cities.html.

Erica Bizzell, "Microbial Ninja Warriors: Bacterial Immune Evasion," American Society for Microbiology, December 2018, https://asm.org/Articles/2018/December/Microbial-Ninja -Warriors-Bacterial-Immune-Evasion.

Nick Carne, "Bacteria with Robust Memories," *Cosmos*, April 28, 2020, https://cosmosmagazine .com/biology/bacteria-with-robust-memories.

Sophia Chen, "These Bacteria Ate Their Way through a Really Tricky Maze," *Wired*, June 9, 2020, https://www.wired.com/story/these-bacteria-ate-their-way-through-a-really-tricky-maze/.

Aimee Cunningham, "Human Skin Bacteria Have Cancer-Fighting Powers," *ScienceNews*, February 28, 2018, https://www.sciencenews.org/article/human-skin-bacteria-have-cancer -fighting-powers.

Alison Escalante, "Scientists Just Brought Us One Step Closer to a Living Computer," *Forbes*, April 29, 2020, https://www.forbes.com/sites/alisonescalante/2020/04/29/scientists-just -brought-us-one-step-closer-to-a-living-computer/?sh=7d546b511f03.

Robert Labayen, "The Mind of a Virus (Why We Are All Connected for a Reason)," ABS-CBN News, April 30, 2020, https://news.abs-cbn.com/blogs/opinions/04/30/20/the-mind-of-a -virus-why-we-are-all-connected-for-a-reason.

Michael Marshall, "Why Microbes Are Smarter Than You Thought," *New Scientist*, June 30, 2009, https://www.newscientist.com/article/dn17390-why-microbes-are-smarter-than-you -thought/.

Abby Ohlheiser, "Maybe It's Time to Retire the Idea of 'Going Viral,' " *MIT Technology Review*, May 17, 2020, https://www.technologyreview.com/2020/05/17/1001809/maybe-its-time-to -retire-the-idea-of-going-viral/.

Elizabeth Pennisi, "The Secret Language of Bacteria," *New Scientist*, September 16, 1995, https://www.newscientist.com/article/mg14719953-500-the-secret-language-of-bacteria/.

Dayana B. Rivadeneira et al., "Oncolytic Viruses Engineered to Enforce Leptin Expression Reprogram Tumor-Infiltrating T Cell Metabolism and Promote Tumor Clearance," *Immunity* 51, no. 3 (September 17, 2019) https://pubmed.ncbi.nlm.nih.gov/31471106/.

Karin Sauer, "Unlocking the Secrets of Bacterial Biofilms—to Use against Them," *Conversation*, May 31, 2016, https://theconversation.com/unlocking-the-secrets-of-bacterial-biofilms-to -use-against-them-59148.

Eric Sawyer, "Editing Genomes with the Bacterial Immune System," Scitable, February 9, 2013, https://www.nature.com/scitable/blog/bio2.0/editing_genomes_with_the_bacterial/.

Derek J. Skillings, "I, Holobiont: Are You and Your Microbes a Community or a Single Entity?"

Aeon, September 26, 2018, https://aeon.co/ideas/i-holobiont-are-you-and-your-microbes-a-community-or-a-single-entity.

Nitin Sreedhar, "When a Virus Jumps: Of Man, Microbes and Pandemics," Livemint.com, April 25, 2020, https://www.livemint.com/mint-lounge/features/when-a-virus-jumps-of-man-microbes-and-pandemics-11587735889206.html.

Jennifer Tsang, "How Can a Slime Mold Solve a Maze? The Physiology Course Is Finding Out," Marine Biological Laboratory, July 24, 2017, http://social.mbl.edu/how-can-a-slime-mold-solve-a-maze-the-physiology-course-is-finding-out.

University of Göttingen, "Bacteria Loop-the-Loop," EurekAlert!, February 27, 2020, https://eurekalert.org/pub_releases/2020-02/uog-bl022720.php.

Jake L. Weissman, Hao H. Yiu, and Philip L. F. Johnson, "What Bacteria Do When They Get Sick," *Frontiers for Young Minds*, July 24, 2019, https://kids.frontiersin.org/article/10.3389/frym.2019.00102.

02. We Are Family

Marwa Azab, PhD, "Gut Bacteria Can Influence Your Mood, Thoughts, and Brain," *Psychology Today*, August 7, 2019, https://www.psychologytoday.com/us/blog/neuroscience-in-every-day-life/201908/gut-bacteria-can-influence-your-mood-thoughts-and-brain.

Martin J. Blaser, "Human Health Is in the Hands of Bacteria," *Time*, October 24, 2019, https://time.com/5709381/human-health-bacteria/.

Dr. Siri Carpenter, "That Gut Feeling," American Psychological Association, September 2012, https://www.apa.org/monitor/2012/09/gut-feeling.

Children's Hospital of Philadelphia, "Researchers Get Important Glimpse into Microbiome Development in Early Life," EurekAlert!, April 16, 2020, https://eurekalert.org/pub_releases/2020-04/chop-rgi041620.php.

Kara Fitzgerald, "Skin Microbiome 101: How to Nurture Good Bacteria & Glowing Skin," MindBodyGreen.com, June 30, 2020, https://www.mindbodygreen.com/0-23996/your-skin-microbiome-why-its-essential-for-a-healthy-glow.html.

GMFH Editing Team, "Your Microbiome Is Like a Unique Fingerprint," Gut Microbiota for Health, July 1, 2015, https://www.gutmicrobiotaforhealth.com/your-microbiome-is-like-a-unique-fingerprint/.

Elizabeth Gulino, "You've Heard of Probiotics & Prebiotics—But What Are Postbiotics?" Refinery 29.com, February 5, 2020, https://www.refinery29.com/en-us/what-are-postbiotics.

Mindy Hermann, "Discover the World of Postbiotics," *Today's Dietitian*, June/July 2020, https://www.todaysdietitian.com/newarchives/JJ20p20.shtml.

Amy Loughman and Heidi Staudacher, "Should I Test My Gut Microbes to Improve My Health?" *Conversation*, March 15, 2020, https://theconversation.com/should-i-test-my-gut-microbes-to-improve-my-health-131216.

Lisa Marshall, "Baby Microbiome: Nurturing Your Baby's Healthy Bacteria," WebMD, https://www.webmd.com/parenting/features/baby-microbiome#1.

Leslie Nemo, "Scientists Can Tell How Old You Are Based on Your Skin Microbiome," *Discover*, February 11, 2020, https://www.discovermagazine.com/health/scientists-can-tell-how-old-you-are-based-on-your-skin-microbiome?utm_source=email&utm_medium=abstract&utm_campaign=mar20_subs.

Anahad O'Connor, "Sucking Your Child's Pacifier Clean May Have Benefits," *Well* (blog), *New York Times*, May 6, 2013, https://well.blogs.nytimes.com/2013/05/06/why-dirty-pacifiers-may-be-your-childs-friend/.

Lois Parshley, "Microbiome Science Could Bring a Revolution in Medical Care," NBCNews.com, January 26, 2017, https://www.nbcnews.com/mach/science/microbiome-science-could-bring-revolution-medical-care-n710861.

Julia Peterman, "Beyond Probiotics: Pre-, Syn-, Post-, and Psycho-," *WholeFoods Magazine*, May 22, 2019, https://wholefoodsmagazine.com/front-page/beyond-probiotics-pre-syn-post-and-psycho/.

Alina Petre, "The Microbiome Diet: Can It Restore Your Gut Health?" Healthline, January 22, 2019, https://www.healthline.com/nutrition/microbiome-diet.

Ruairi Robertson, "How Your Gut Bacteria Can Influence Your Weight," Healthline, February 13, 2018, https://www.healthline.com/nutrition/gut-bacteria-and-weight.

Ana Sandoiu, "Can Gut Bacteria Predict Your Personality?" Medical News Today, February 17,

2020, https://www.medicalnewstoday.com/articles/can-your-gut-bacteria-predict-your
-personality.

Justin Sonnenburg and Erica Sonnenburg, "Gut Feelings—The 'Second Brain' in Our Gastrointestinal Systems [Excerpt]," *Scientific American*, May 1, 2015, www.scientificamerican.com
/article/gut-feelings-the-second-brain-in-our-gastrointestinal-systems-excerpt/.

Elizabeth Svoboda, "Gut Bacteria's Role in Anxiety and Depression: It's Not Just in Your Head,"
Discover, October 4, 2020, https://www.discovermagazine.com/mind/gut-bacterias-role-in
-anxiety-and-depression-its-not-just-in-your-head.

Clare Wilson, "Babies Are Being Fed Mother's Poo in Effort to Boost Gut Bacteria," *New Scientist*,
October 1, 2020, https://www.newscientist.com/article/2255964-babies-are-being-fed
-mothers-poo-in-effort-to-boost-gut-bacteria/#ixzz6aNmCSoXu.

Colleen Zacharyczuk, "Human Microbiome Project May Hold Promise for Future," Healio.com,
May 30, 2012, https://www.healio.com/news/pediatrics/20120611/human-microbiome
-project-may-hold-promise-for-future.

03. Rewilding Our Urban Microbiome

"About the Home Microbiome Project," Home Microbiome, https://homemicrobiome.com/the
-home-microbiome-study/.

Diana Budds, "How Microbes 'Designed' New York," Curbed New York, October 2, 2018, https://
ny.curbed.com/2018/10/2/17925378/nyc-museum-exhibit-germs-public-housing.

Cell Press, "Microbes in the Hong Kong Subway System Mix Together by Evening Rush Hour," Phys
.org, July 31, 2018, https://phys.org/news/2018-07-microbes-hong-kong-subway-evening.html.

Michelle Cohen, "An Eight-Story Monarch Butterfly Sanctuary May Be the Façade of a New
Nolita Building," 6sqft.com, March 2, 2020, https://www.6sqft.com/an-eight-story-monarch
-butterfly-sanctuary-may-be-the-facade-of-a-new-nolita-building/.

Richard Conniff, "Take a Deep Breath and Say Hi to Your Exposome," *Scientific American*, September 28, 2018, https://www.scientificamerican.com/article/take-a-deep-breath-and-say
-hi-to-your-exposome/.

Ned Dodington, "Buildings + Germs," Expanded Environment, July 13, 2020, http://www
.expandedenvironment.org/buildings-vs-germs/.

Veronique Greenwood, "The Mysterious Microbes in the Sky," *Atlantic*, June 29, 2018, https://
www.theatlantic.com/science/archive/2018/06/microbes-can-change-the-temperature
/564092/.

Lakshmi Iyengar, "You Have a Microbial Cloud!" *Yale Scientific*, February 3, 2016, https://www
.yalescientific.org/2016/02/you-have-a-microbial-cloud/.

Gideon Lasco, "Could COVID-19 Permanently Change Hand Hygiene?" *Sapiens*, April 8, 2020,
https://www.sapiens.org/biology/hand-hygiene-covid-19/.

Lucie Levine, "The Lower East Side's Forgotten Lung Block: The Italian Community Lost to 'Slum
Clearance,' " 6sqft.com, April 25, 2019, https://www.6sqft.com/the-lower-east-sides-forgotten
-lung-block-the-italian-community-lost-to-slum-clearance/.

Daphne Miller, "Uncovering How Microbes in the Soil Influence Our Health and Our Food,"
Washington Post, September 29, 2019, https://www.washingtonpost.com/science/uncovering
-how-microbes-in-the-soil-influence-our-health-and-our-food/2019/09/27/81634f54-a4ba-11
e9-bd56-eac6bb02d01d_story.html.

"Researchers Produce First Map of New York City Subway System Microbes," Weill Cornell Medicine, February 5, 2015, https://news.weill.cornell.edu/news/2015/02/researchers-produce
-first-map-of-new-york-city-subway-system-microbes-christopher-mason.

Jake M. Robinson, "Biodiversity Loss Could Be Making Us Sick—Here's Why," *Conversation*, August 4, 2020, https://theconversation.com/biodiversity-loss-could-be-making-us-sick-heres
-why-143627.

Predrag Slijepcevic, "Bacteria and Viruses Are Travelling the World on Highways in the Sky,"
Conversation, July 21, 2020, https://theconversation.com/bacteria-and-viruses-are-travelling
-the-world-on-highways-in-the-sky-142854.

Chelsea Whyte, "When You Ride the Subway You Share Bacteria with Everyone in Your City,"
New Scientist, July 31, 2018, https://www.newscientist.com/article/2175556-when-you-ride
-the-subway-you-share-bacteria-with-everyone-in-your-city/.

"Your Personal Microbial Cloud," Institute for Health in the Built Environment, University of
Oregon, https://buildhealth.uoregon.edu/feature-research-your-personal-microbial-cloud/.

04. Here Come the Biotopians

Mihai Andrei, "While the World Is Looking at AI, It's Biotech That Could End Up Changing the World," ZME Science, May 29, 2020, https://www.zmescience.com/science/biotech-change-world-23052020/.

Ruben Baart, "Suzanne Lee Wants to Live in a Sustainable Material World. Built with Biology, Not Oil," Next Nature Network, May 28, 2020, https://nextnature.net/story/2020/interview-suzanne-lee-2.

"Best Fermented Foods from Around the World," Happy Belly Fish, February 1, 2020, https://happybellyfish.com/2020/02/01/best-fermented-foods-from-around-the-world/.

"Biodesign Challenge," https://biodesignchallenge.org/.

Kristen Catalano, "Breaking the Mold: How This Vegan Dog Food Company Is Taking on the Big Brands," This Dog's Life, October 3, 2019, https://www.thisdogslife.co/breaking-the-mold-how-this-vegan-dog-food-company-is-taking-on-the-big-brands/.

Charles Choi, "This Regenerative Building Material Is Made from Sand and Bacteria," *Discover*, January 21, 2020, https://www.discovermagazine.com/the-sciences/this-regenerative-building-material-is-made-from-sand-and-bacteria.

Freeman Dyson, "Our Biotech Future," *New York Review of Books*, July 19, 2007, https://www.nybooks.com/articles/2007/07/19/our-biotech-future/.

Clara Rodríguez Fernández, "CRISPR-Cas9: The Gene Editing Tool Changing the World," Labiotech.eu, September 7, 2020, https://www.labiotech.eu/in-depth/crispr-cas9-review-gene-editing-tool/.

Lorin Fries, "The Foods of Tomorrow: How Biotechnology Is Changing What We Eat," *Forbes*, June 19, 2019, https://www.forbes.com/sites/lorinfries/2019/06/19/the-foods-of-tomorrow-how-biotechnology-is-changing-what-we-eat/?sh=77146edc1235.

Vicky Gan, "How Germs Might Shape the Future of Architecture," CityLab, May 11, 2015, https://www.citylab.com/life/2015/05/how-germs-might-shape-the-future-of-architecture/392783/.

Ginkgo Bioworks, *Grow*, GrowbyGinkgo.com, https://www.growbyginkgo.com/.

"IndieBio: Creating the Future of Food," Cell.ag, December 19, 2018, https://www.cell.ag/blog/indiebio-creating-future-of-food.

Brooke Roberts-Islam, "Could This Innovation Be an Answer to Fashion's Plastic Problem?" *Forbes*, November 30, 2020, https://www.forbes.com/sites/brookerobertsislam/2020/11/30/could-this-innovation-be-an-answer-to-fashions-plastic-problem/.

Muchaneta Kapfunde, "Suzanne Lee's Biofabricate, Growing a Better Future," FashNerd.com, 2017, https://fashnerd.com/2017/12/biofabrication-suzannelee-fashion-tech-sustainability.

Gareth John Macdonald, "Biomanufacturing Makes Sense of the Industry 4.0 Concept," Genetic Engineering & Biology News, August 3, 2020, https://www.genengnews.com/insights/biomanufacturing-makes-sense-of-the-industry-4-0-concept/.

Anna Marks, "Art Meet Bio, Bio Meet Art: Designing a Sustainable Future," Synbiobeta.com, September 29, 2019, https://synbiobeta.com/art-meet-bio-bio-meet-art-designing-a-sustainable-future/.

Garrett Parker, "What Is Biomanufacturing and How Will It Change the World?" *Money Inc.*, 2015, https://moneyinc.com/biomanufacturing/.

Emily Waltz, "Appetite Grows for Biotech Foods with Health Benefits," *Nature*, April 29, 2019, https://www.nature.com/articles/d41587-019-00012-9.

Jonathan Webb, "Oldest Human Faeces Show Neanderthals Ate Vegetables," BBC News, June 26, 2014, https://www.bbc.com/news/science-environment-27981702.

05. Ancient High Technology

Jef Akst, "The Influence of Soil on Immune Health," *Scientist*, January 8, 2020, https://www.the-scientist.com/news-opinion/the-influence-of-soil-on-human-health-66885.

Paul Scott Anderson, "These Bacteria Eat and Breathe Electricity," Earthsky.org, March 16, 2019, https://earthsky.org/earth/scientists-study-bacteria-that-eat-and-breathe-electricity#.

Colin Barras, "Bacteria: The Miracle Microbes That Could Fix Planet," *Science Focus*, September 2, 2020, https://www.sciencefocus.com/science/bacteria-the-miracle-microbes-that-could-fix-planet.

Elle Bethune, "Electricity from Bacteria: The Future of Clean Energy?" Eu:Sci, December 8, 2020, http://eusci.org.uk/2020/12/08/electricity-from-bacteria-the-future-of-clean-energy/.

Leigh Krietsch Boerner, "Biocellection's Miranda Wang and Jeanny Yao Aim to Make Treasure

Out of Plastic Trash," *Chemical & Engineering News*, March 8, 2020, https://cen.acs.org
/environment/recycling/BioCellection-Miranda-Wang-and-Jeanny-Yao-aim-to-make-treasure
-out-of-plastic-trash/98/i9.

Jordan Davidson, "Scientists Find Bacteria That Eats Plastic," EcoWatch, March 27, 2020, https://
www.ecowatch.com/scientists-find-bacteria-that-eats-plastic-2645582039.html.

Jason Dorrier, "AI Just Discovered a New Antibiotic to Kill the World's Nastiest Bacteria," Singu-
larity Hub, February 23, 2020, https://singularityhub.com/2020/02/23/for-the-first-time
-ai-finds-a-new-antibiotic-to-kill-the-worlds-nastiest-bacteria/.

DTE Staff, "New Bacteria Will Help Fight Climate Change, Soil Pollutants: Researchers," Down
ToEarth.org, February 21, 2020, https://www.downtoearth.org.in/news/climate-change/new
-bacteria-will-help-fight-climate-change-soil-pollutants-researchers-69407#.

Ivy Engel, "How a Bacteria from Yellowstone Is Helping Fight Coronavirus," Wyoming Public
Media, April 20, 2020, https://www.wyomingpublicmedia.org/post/how-bacteria-yellow
stone-helping-fight-coronavirus#stream/0.

Maya Wei-Haas, "Key Ingredient in Coronavirus Tests Comes from Yellowstone's Lakes," *National
Geographic*, March 31, 2020, https://www.nationalgeographic.com/science/2020/03/key
-ingredient-in-coronavirus-tests-comes-from-yellowstone/.

Olga Kuchment, "How Soil Microbes Help Plants Resist Disease," AgriLife Today, April 6, 2020,
https://agrilifetoday.tamu.edu/2020/04/06/how-soil-microbes-help-plants-resist-disease/.

Derek Lovley, "New Green Technology from UMass Amherst Generates Electricity 'Out of Thin
Air,' " University of Massachusetts Amherst, February 17, 2020, https://www.umass.edu
/newsoffice/article/new-green-technology-umass-Amherst.

Joe McKendrick, "A Bio Internet of Things? Hold That Thought," RTInsights.com, December 6,
2019, https://www.rtinsights.com/bio-internet-of-things-hold-that-thought/.

National University of Singapore, "Bacteria Could Make Cheaper Biofuel Without So Much Land,"
Futurity, May 14, 2018, https://www.futurity.org/mushroom-waste-bacteria-biofuel-1756722.

Elizabeth Pennisi, " 'Electric Mud' Teems with New, Mysterious Bacteria," *Science*, August 19,
2020, https://www.sciencemag.org/news/2020/08/electric-mud-teems-new-mysterious
-bacteria#.

Rachel Ross, "The Science behind Composting," Live Science, September 12, 2018, https://
www.livescience.com/63559-composting.html.

Alexander Lipinski, "What Bacteria Can Teach Us about Efficient Methane Production," Bioeconomy BW, May 14, 2012, https://www.biooekonomie-bw.de/en/articles/news/what
-bacteria-can-teach-us-about-efficient-methane-production.

Graham Templeton, "How MIT's New Biological 'Computer' Works, and What It Could Do in the
Future," ExtremeTech, July 25, 2016, https://www.extremetech.com/extreme/232190-how
-mits-new-biological-computer-works-and-what-it-could-do-in-the-future.

Anne Trafton, "Artificial Intelligence Yields New Antibiotic," *MIT News*, February 20, 2020, https://
news.mit.edu/2020/artificial-intelligence-identifies-new-antibiotic-0220#.

University of Manchester, "Biofuels Could Be Made from Bacteria That Grow in Seawater,"
Biomass Magazine, October 31, 2019, http://biomassmagazine.com/articles/16583/biofuels
-could-be-made-from-bacteria-that-grow-in-seawater.

Jonathan Watts, "Earth's Depths Are Teeming with Otherworldly Microbes," *Wired*, December
10, 2018, https://www.wired.com/story/deep-carbon-observatory-subterranean-ecosystems/.

Joseph Wolkin, "Amoeba Defeats Challenging 'Traveling Salesman Problem,' " Interesting
Engineering, December 24, 2018, https://interestingengineering.com/amoeba-defeats
-challenging-traveling-salesman-problem.

06. Micro-Nauts

Jamie Carter, "What Is 'Panspermia'? New Evidence for the Wild Theory That Says We Could
All Be Space Aliens," *Forbes*, August 26, 2020, https://www.forbes.com/sites/jamiecarter
europe/2020/08/26/what-is-panspermia-new-evidence-for-the-wild-theory-that-says-we
-could-all-be-space-aliens/?sh=3bd6aa686543.

Adam Frank, "What If Life on Earth Didn't Start on Earth?" NPR, December 15, 2017, https://
www.npr.org/sections/13.7/2017/12/15/571122951/what-if-life-on-earth-didn-t-start-on
-earth.

"Growing Microbes in Space," Biocompare, April 10, 2020, https://www.biocompare.com/Life
-Science-News/563059-Growing-Microbes-in-Space/.

Loren Grush, "Discovery of Noxious Gas on Venus Could Be a Sign of Life," *Verge*, September 14, 2020, https://www.theverge.com/21428796/venus-gas-life-sign-discovery-phosphine-biosignature.

Katie Hunt, "Bacteria from Earth Could Potentially Be Used to Mine on the Moon or Mars," CNN, November 10, 2020, https://www.cnn.com/2020/11/10/world/microbes-mining-in-space-scn/index.html.

Alla Katsnelson, "Microbes Mine Rare Earths in Space," C&EN, November 16, 2020, https://pubs.acs.org/doi/10.1021/cen-09844-scicon5.

Marc Kaufman, "In Search of Panspermia," Astrobiology at NASA, January 6, 2017, https://astrobiology.nasa.gov/news/in-search-of-panspermia/.

Lauren Kent, "The European Space Agency Plans to Start Mining for Natural Resources on the Moon," CNN, January 22, 2019, https://www.cnn.com/2019/01/22/europe/mining-on-moon-trnd/index.html.

Jennifer Leman, "Even More Evidence Points to 'Water Bodies' Under the Surface of Mars," *Popular Mechanics*, September 30, 2020, https://www.popularmechanics.com/space/moon-mars/a34224730/liquid-water-lakes-mars/#.

Rafi Letzter, "An Ancient Virus May Be Responsible for Human Consciousness," Live Science, February 2, 2018, https://www.livescience.com/61627-ancient-virus-brain.html.

Gilbert V. Levin, "I'm Convinced We Found Evidence of Life on Mars in the 1970s," *Scientific American*, October 10, 2019, https://blogs.scientificamerican.com/observations/im-convinced-we-found-evidence-of-life-on-mars-in-the-1970s/.

Dan Nosowitz, "Good Old-Fashioned Earth Soil Will Be Heading to Space for the First Time," Modern Farmer, September 25, 2020, https://modernfarmer.com/2020/09/good-old-fashioned-earth-soil-will-be-heading-to-space-for-the-first-time/.

Daniel Oberhaus, "A Ball of Bacteria Survived for 3 Years . . . in Space!" *Wired*, August 26, 2020, https://www.wired.com/story/a-ball-of-bacteria-survived-for-3-years-in-space/.

Daniel Oberhaus, "A Crashed Israeli Lunar Lander Spilled Tardigrades on the Moon," *Wired*, August 5, 2019, https://www.wired.com/story/a-crashed-israeli-lunar-lander-spilled-tardigrades-on-the-moon/?verso=true.

Brian Resnick, "Tardigrades, the Toughest Animals on Earth, Have Crash-Landed on the Moon," *Vox*, August 6, 2019, https://www.vox.com/science-and-health/2019/8/6/20756844/tardigrade-moon-beresheet-arch-mission.

Jim Robbins, "Trillions upon Trillions of Viruses Fall from the Sky Each Day," *New York Times*, April 13, 2018, https://www.nytimes.com/2018/04/13/science/virosphere-evolution.html.

Kara Rogers, "Abiogenesis," *Encyclopedia Britannica*, September 26, 2018, https://www.britannica.com/science/abiogenesis.

David Rothery, "New Water Plumes from Jupiter's Moon Europa Raise Hopes of Detecting Microbial Life," *Conversation*, September 26, 2016, https://theconversation.com/new-water-plumes-from-jupiters-moon-europa-raise-hopes-of-detecting-microbial-life-66019.

Jillian Scudder, "Could Life Exist in a Star's Atmosphere?" *Forbes*, December 8, 2015, https://www.forbes.com/sites/startswithabang/2015/12/08/astroquizzical-life-stellar-atmospheres/?sh=7bafdcbc6617.

Ivy Shih, "Bacteria Found to Thrive Better in Space Than on Earth," *Conversation*, March 24, 2016, https://theconversation.com/bacteria-found-to-thrive-better-in-space-than-on-earth-56740.

Abigail Tabor, "Astronauts Leave 'Microbial Fingerprint' on Space Station," NASA, March 2, 2021, https://www.nasa.gov/feature/ames/astronauts-leave-microbial-fingerprint-on-space-station.

Additional Reading

Martin J. Blaser, *Missing Microbes: How the Overuse of Antibiotics Is Fueling Our Modern Plagues* (New York: Henry Holt and Company, 2014).

Po Bronson & Arvind Gupta, *Decoding the World: A Roadmap for the Questioner* (New York: Twelve Books, 2020).

Robynne Chutkan, *The Microbiome Solution: A Radical New Way to Heal Your Body from the Inside Out* (New York: Avery, 2015).

Jared Diamond, *Guns, Germs, and Steel: The Fates of Human Societies* (New York: W. W. Norton & Company, 1997).

Jennifer A. Doudna and Samuel H. Sternberg, *A Crack in Creation: Gene Editing and the Unthinkable Power to Control Evolution* (Boston: Mariner Books, 2018).

Paul G. Falkowski, *Life's Engines: How Microbes Made Earth Habitable* (Princeton, NJ: Princeton University Press, 2015).

Sir Fred Hoyle and Chandra Wickramasinghe, *Evolution from Space: A Theory of Cosmic Creationism* (New York: Touchstone, 1981).

——, *Lifecloud: The Origin of Life in the Universe* (New York: Harper & Row, 1978).

Raphael Kellman, MD, *The Microbiome Diet: The Scientifically Proven Way to Restore Your Gut Health and Achieve Permanent Weight Loss* (New York: Da Capo Lifelong Books, 2014).

Suzanne Lee, *Fashioning the Future: Tomorrow's Wardrobe* (New York: Thames & Hudson, 2005).

Iris Lewandowski, ed., *Bioeconomy: Shaping the Transition to a Sustainable, Biobased Economy* (New York: Springer, 2018).

Anne Maczulak, *Allies and Enemies: How the World Depends on Bacteria* (Upper Saddle River, NJ: FT Press, 2010).

Matt McCarthy, MD, *Superbugs: The Race to Stop an Epidemic* (New York: Avery, 2019).

David Julian McClements, *Future Foods: How Modern Science Is Transforming the Way We Eat* (New York: Springer, 2019).

Michael B. A. Oldstone, *Viruses, Plagues & History: Past, Present, and Future* (New York: Oxford University Press, 2010).

Chandra Wickramasingh, PhD; Kamala Wickramasinghe, MA; and Gensuke Tokoro, *Our Cosmic Ancestry in the Stars: The Panspermia Revolution and the Origins of Humanity* (Rochester, VT: Bear & Company, 2019).

Ed Yong, *I Contain Multitudes: The Microbes within Us and a Grander View of Life* (New York: Ecco Press, 2016).

People We Noted

Thomas Abraham, adjunct associate professor at the University of Hong Kong's Journalism and Media Studies Centre, https://jmsc.hku.hk/people/thomas-abraham/

Gladys Alexandre, associate professor of biochemistry, https://bcmb.utk.edu/people/faculty/gladys-alexandre/

Alejandro Amezquita, future bio-based ingredients R&D director, Unilever F&R, https://www.linkedin.com/in/amezquita

Robert Austin, physicist at Princeton University, https://phy.princeton.edu/people/robert-austin

Bonnie L. Bassler, Squibb Professor in Molecular Biology, https://scholar.princeton.edu/basslerlab/home

Alan L. Bean, Apollo 12 astronaut, https://www.nasa.gov/feature/alan-bean/

Ryan Bethencourt, venture capitalist, https://www.linkedin.com/in/bethencourt/

Martin J. Blaser, professor of medicine and microbiology at Rutgers University, https://cabm.rutgers.edu/person/martin-j-blaser

Thomas Brock, director of the Community College Research Center, https://ccrc.tc.columbia.edu/person/thomas-brock.html

Natsai Audrey Chieza, London-based designer, https://www.natsaiaudrey.co.uk/

George Church, PhD, geneticist, https://wyss.harvard.edu/team/core-faculty/george-church/

Rita R. Colwell, environmental microbiologist, https://www.umiacs.umd.edu/people/rita-colwell

Pete Conrad, Apollo 12 astronaut, https://en.wikipedia.org/wiki/Pete_Conrad

John F. Cryan, neuropharmacologist at University College Cork, https://apc.ucc.ie/john_cryan/

Charles Darwin, British naturalist, https://www.biography.com/scientist/charles-darwin

Richard Dawkins, evolutionary biologist, https://www.edge.org/memberbio/richard_dawkins

Floyd Dewhirst, microbiologist, https://www.forsyth.org/scientists/floyd-dewhirst/

Steven D'Hondt, professor of oceanography at the University of Rhode Island, https://web.uri.edu/gso/meet/steven-dhondt/

Ted Dinan, psychiatrist at University College Cork, https://apc.ucc.ie/ted_dinan/

Rob Dunn, biologist at North Carolina State University, https://cals.ncsu.edu/applied-ecology/people/rob-dunn/

Freeman J. Dyson, physicist and theorist, https://en.wikipedia.org/wiki/Freeman_Dyson

Susan Erdman, principal research scientist at MIT, http://erdmanlab.us/about.php

Johann August Ephraim Goeze, German zoologist, https://en.wikipedia.org/wiki/Johann_August_Ephraim_Goeze

David H. Grinspoon, astrobiologist, https://www.psi.edu/about/staffpage/grinspoon

Daniel Grushkin, founder of Biodesign Challenge, https://biodesignchallenge.org/our-team

Sir Fred Hoyle, cosmologist at the University of Cambridge, https://en.wikipedia.org/wiki/Fred_Hoyle

Naveen Jain, CEO of Viome, https://www.linkedin.com/in/naveenjainintelius/

Katerina Johnson, experimental psychologist at Oxford University, https://www.ox.ac.uk/news-and-events/find-an-expert/katerina-johnson

Raphael Kellman, board-certified physician, https://kellmancenter.com/

Raphael Kim, Queen Mary University of London, https://www.linkedin.com/in/raphael-kim-59b81b25/?originalSubdomain=uk

Joshua Lederberg, molecular biologist, https://en.wikipedia.org/wiki/Joshua_Lederberg

Jay T. Lennon, professor of biology at Indiana University Bloomington, https://biology.indiana.edu/about/faculty/lennon-jay.html

Sanjay Limaye, planetary scientist, https://experts.news.wisc.edu/experts/sanjay-limaye

Kenneth J. Locey, faculty member at Diné College, https://kenlocey.weecology.org/

Richard Losick, American molecular biologist, https://scholar.harvard.edu/rlosick/home

Derek Lovley, microbiologist at the University of Massachusetts Amherst, https://www.micro.umass.edu/faculty-and-research/derek-lovley

Anne A. Madden, PhD, microbiologist, https://www.linkedin.com/in/anne-a-madden-ph-d/

Ioana Man, multidisciplinary designer, https://www.ioanaman.com/about

Lynn Margulis, biologist, https://en.wikipedia.org/wiki/Lynn_Margulis

Dr. William F. Martin, evolutionary biologist, https://www.molevol.hhu.de/en/prof-dr-william-f-martin

Dorothy Matthews, professor at Russell Sage College, https://www.linkedin.com/in/dorothy-matthews-2b462120/

Stella McCartney, designer, https://www.stellamccartney.com/experience/en/about-stella/

Arianna Miano, UC San Diego bioengineering PhD student, https://www.linkedin.com/in/arianna-miano-b41339131/

Harold J. Morowitz, biophysicist, https://en.wikipedia.org/wiki/Harold_J._Morowitz

Kary B. Mullis, https://www.nobelprize.org/prizes/chemistry/1993/mullis/facts/

Jason Peters, assistant professor in the University of Wisconsin-Madison School of Pharmacy's Pharmaceutical Sciences Division, https://energy.wisc.edu/about/energy-experts/jason-peters

Dr. Ilka Bischofs-Pfeifer, principal investigator of the research team at the Max Planck Institute, https://www.mpi-marburg.mpg.de/654781/Ilka-Bischofs-Pfeifer.html

Trung T. Phan, CFA and writer at the *Hustle*, https://trungtphan.com/

Dr. Stefan Poslad, senior lecturer at Queen Mary University of London, http://eecs.qmul.ac.uk/profiles/posladstefan.html

Jake M. Robinson, ecologist and PhD researcher, Department of Landscape, University of Sheffield, https://theconversation.com/profiles/jake-m-robinson-1139158

Carl Sagan, astronomer, https://www.planetary.org/profiles/carl-sagan

Karin Sauer, professor of biological sciences, Binghamton University, https://www.binghamton.edu/biology/people/profile.html?id=ksauer

Nigel Scrutton, University of Manchester, https://www.research.manchester.ac.uk/portal/nigel.scrutton.html

Julia Segre, PhD, scientist, https://irp.nih.gov/pi/julie-segre

Derek Skillings, biologist and philosopher of science, https://aeon.co/users/derek-skillings

Wil V. Srubar, assistant professor in the Department of Civil, Environmental and Architectural Engineering (CEAE), University of Colorado, https://www.colorado.edu/ceae/wil-v-srubar

Gürol Süel, head of the laboratory at University of California, San Diego, https://biology.ucsd.edu/research/faculty/gsuel

Vernor Vinge, retired San Diego State University professor of mathematics, https://stage.edge.org/memberbio/vernor_vinge

Miranda Wang, cofounder of Novoloop, https://www.ioes.ucla.edu/person/miranda-wang/

Chandra Wickramasinghe, astrobiologist, https://www.bl.uk/voices-of-science/interviewees/chandra-wickramasinghe

Dan Widmaier, founder and CEO, Bolt Threads, https://www.linkedin.com/in/dan-widmaier-1760601/

Jeanny Yao, cofounder of Novoloop, https://www.linkedin.com/in/jeanny-yao/

Jun Yao, electrical engineer at the University of Massachusetts Amherst, https://www.umass.edu/ials/people/jun-yao

Alex Zhavoronkov, founder of InSilicoMedicine, https://insilico.com/

Organizations

Algalife, https://www.alga-life.com/
Algenuity, https://www.algenuity.com/
American College of Obstetricians and Gynecologists (ACOG), https://www.acog.org/
American Pet Products Association (APPA), https://www.americanpetproducts.org/
American Society for Microbiology, https://asm.org/
Amino Labs, https://amino.bio/
Argonne National Laboratory, https://www.anl.gov/
ArianeGroup, https://www.ariane.group/en/
Arizona State University, https://www.asu.edu/
AsimicA, http://www.asimica.com/
Ben-Gurion University of the Negev, Israel, https://in.bgu.ac.il/en/pages/default.aspx
Berkeley Lab at the University of California, Berkeley, https://www.lbl.gov/
Beyond Meat, https://www.beyondmeat.com/
Biodesign Challenge, https://biodesignchallenge.org/
Biodesigned, https://www.biodesigned.org/
Bioscience Institute, https://bioinst.com/en/
Bolt Threads, https://boltthreads.com/
California Institute of Technology, https://www.caltech.edu/
Cambrian Innovation, Massachusetts, https://cambrianinnovation.com/
Carbios, https://carbios.fr/en/
Census of Marine Life, http://www.coml.org/
Center for Genome Sciences & Systems Biology, Washington University School of Medicine,
    St. Louis, Missouri, https://genomesciences.wustl.edu/
Children's Hospital of Philadelphia (CHOP), https://www.chop.edu/
CNN, https://www.cnn.com/
Commercial Crew Program, SpaceX and NASA, https://www.nasa.gov/exploration/commercial
    /crew/index.html
Community of Microbes at Cooper Hewitt, https://cooper.edu/events-and-exhibitions/exhibi
    tions/community-microbes
Cooper Union Colonnade at Cooper Union, New York City, https://cooper.edu/art/student
    exhibitions
Cornell University College of Agriculture and Life Sciences School of Integrative Plant Science,
    https://cals.cornell.edu/school-integrative-plant-science
Cronos Group, https://www.thecronosgroup.com/
Culture Biosciences, https://www.culturebiosciences.com/
Day Two, https://www.daytwo.com/
Diné College, https://www.dinecollege.edu/
European Algae Biomass Association (EABA), https://www.eaba-association.org/en
European Space Agency, https://www.esa.int/
Faber Futures, https://faberfutures.com/
Genspace, https://www.genspace.org/
George Washington University, https://www.gwu.edu/
Germ City: Microbes and the Metropolis, Museum of the City of New York, https://www.mcny
    .org/exhibition/germ-city
Ginkgo Bioworks, https://www.growbyginkgo.com/
Good Food Institute, https://www.gfi.org/
Harvard Medical School, https://hms.harvard.edu/
Harvard University T. H. Chan School of Public Health, https://www.hsph.harvard.edu/
Heinrich Heine University, Düsseldorf, Germany, https://www.hhu.de/en/
Hoekmine BV, https://www.hoekmine.com/
Holobiome, https://holobiome.org/
Homebiotic, https://homebiotic.com/
Human Microbiome Project, NYU Langone Health, http://gerd.med.nyu.edu/hmp
Humane Society of the United States, https://secure.humanesociety.org/
Imperial College London, https://www.imperial.ac.uk/

Indiana University Bloomington, https://www.indiana.edu/
IndieBio, https://indiebio.co/
Insilico Medicine, https://insilico.com/
Institute for Health in the Built Environment, University of Oregon, https://buildhealth.uoregon
.edu/feature-research-your-personal-microbial-cloud/
International Committee Against Mars Sample Return (ICAMSR), http://www.icamsr.org/
International Council of Nurses (ICNP), https://www.icn.ch/what-we-do/projects/ehealth-icnptm
/about-icnp
International Space Station (ISS), https://www.nasa.gov/mission_pages/station/main/index.html
Italian Space Agency (ISA), https://www.asi.it/en/
Johns Hopkins Medicine, https://www.hopkinsmedicine.org/
Kallyope, https://www.kallyope.com/
Keio University, Tokyo, Japan, https://www.keio.ac.jp/en/
Lawrence Berkeley National Laboratory, https://www.lbl.gov/
Live Science, https://www.livescience.com/
Map My Environment, https://www.mapmyenvironment.com/
MARUM–Center for Marine Environmental Sciences, https://www.marum.de/en/index.html
Materiability Research Group, http://materiability.com/
Max Planck Institute for Marine Microbiology, https://www.mpi-bremen.de/en/Home.html
Max Planck Institute for Terrestrial Microbiology, https://www.mpi-marburg.mpg.de/
Max Planck Institute of Colloids and Interfaces, https://www.mpikg.mpg.de/en
Michigan State University, https://msu.edu/
MIT, https://www.mit.edu/
Moon Juice, https://moonjuice.com/
Museum of Modern Art, https://www.moma.org/
New York Times, https://www.nytimes.com/
North Carolina State University, https://www.ncsu.edu/
Novoloop, https://www.novoloop.com/
NovoNutrients, https://www.novonutrients.com/
Parsons School of Design, https://www.newschool.edu/parsons/
Perfect Day Foods, https://perfectdayfoods.com/
Philips Design, https://www.philips.com/a-w/about/philips-design/design-in-action.html
PILI, https://www.pili.bio/
Plant Based Foods Association, https://plantbasedfoods.org/
Popular Science, https://www.popsci.com/
Princeton University, https://www.princeton.edu/
Reazent, https://reazent.com/
San Diego State University, https://www.sdsu.edu/
School of Architecture, Planning & Landscape, Newcastle University, Newcastle, UK, https://
www.ncl.ac.uk/apl/
Scindo, https://scindo.bio/
Secret Science Club, http://secretscienceclub.blogspot.com/
Solar Foods, https://solarfoods.fi/
Spanish Society for Biochemistry and Molecular Biology, https://www.febs.org/our-members
/the-spanish-society-for-biochemistry-and-molecular-biology/
Stanford Byers Center for Biodesign, https://biodesign.stanford.edu/
Swedish University of Agricultural Sciences, Alnarp, Sweden, https://www.slu.se/en/
Swiss Federal Institute for Forest, Snow and Landscape Research (WSL), https://www.wsl.ch/en
/index.html
TargEDys, https://www.targedys.com/en/
Tel Aviv University, https://english.tau.ac.il/
UN Environment Programme, https://www.unep.org/
Universidad de Granada, Spain, https://www.ugr.es/en/
Universidad de Los Andes, https://uniandes.edu.co/
University of the Arts London, https://www.arts.ac.uk/
University of Bristol, https://bristol.ac.uk/
University of Buckingham, https://www.buckingham.ac.uk/
University of California, Berkeley, https://www.berkeley.edu/

University of California, San Diego, https://ucsd.edu/
University of Cambridge, Cambridge, UK, https://www.cam.ac.uk/
University of Chicago, https://www.uchicago.edu/
University of Colorado Boulder, https://www.colorado.edu/
University of East Anglia Department of Biological Sciences, Norwich, England, https://www
    .uea.ac.uk/about/school-of-biological-sciences
University of Hong Kong, https://www.hku.hk/
University of Oregon, https://www.uoregon.edu/
University of Wisconsin–Madison School of Pharmacy, Pharmaceutical Sciences PhD, https://
    pharmacy.wisc.edu/academics/pharmsci/
Urgent Company, https://theurgentcompany.com/
US Department of Defense, https://dod.defense.gov/
US Department of Health & Human Services, https://www.hhs.gov/
US Food & Drug Administration (FDA), https://www.fda.gov/home
US Geological Survey (USGS), https://www.usgs.gov/
Viome, https://www.viome.com/
VTT Technical Research Centre of Finland, https://www.vttresearch.com/en
Weill Cornell Medicine, https://weill.cornell.edu/
Wellcome Collection, https://wellcomecollection.org/
Wild Earth, https://wildearth.com/
Yellowstone National Park, https://www.nps.gov/yell/index.htm
Yellowstone Volcano Observatory, https://www.usgs.gov/observatories/yellowstone-volcano
    -observatory
Your Super, https://yoursuper.com/pages/our-story
Zymergen, https://www.zymergen.com/

Acknowledgments

Alice in Futureland would not have been possible without the guidance of numerous individuals who in one way or another have contributed their valuable, innovative wisdom, and energy. ☺

First and foremost, we owe our deepest gratitude to our Sputnik Futures and Alice in Futureland colleagues and partners, Lisa, Amy, and Jordan, who have wandered with us through the many years of futures research. Know that we have circled the stars together and have many more laps to go.

Thank you, Luis, for bringing Alice to life.

Special thanks to the like minds at Tiller Press, especially Theresa, who saw the vision for this future series; Sam for his insights; and Hannah for her sharp editing.

And to the many frontier thinkers who have shared their knowledge with enthusiasm, we are gratefully indebted.

To our families who have lived this journey with us, thank you all with love!

J&J

Notes

00. INTRODUCTION

1. "Identifying the Source of the Outbreak," CDC.gov, July 1, 2020, https://www.cdc.gov /coronavirus/2019-ncov/cases-updates/about-epidemiology/identifying-source-outbreak.html.
2. William F. Martin, Madeline C. Weiss, Sinje Neukirchen, Shijulal Nelson-Sathi, and Filipa L. Sousa, "Physiology, Phylogeny, and LUCA," *Microbial Cell* 3, no. 12 (December 2016): 582–87, doi: 10.15698/mic2016.12.545.
3. Ibid.
4. Madeline C. Weiss, Filipa L. Sousa, Natalia Mrnjavac, Sinje Neukirchen, Mayo Roettger, Shijulal Nelson-Sathi, and William F. Martin, "The Physiology and Habitat of the Last Universal Common Ancestor," *Nature Microbiology* 1 (July 2016), doi: 10.1038/nmicrobiol.2016.116.
5. University of Bristol, "A Timescale for the Origin and Evolution of All of Life on Earth," *Science-Daily*, August 20, 2018, https://www.sciencedaily.com/releases/2018/08/180820113113.htm.
6. "HIV/AIDS Glossary: Microorganism," ClinicalInfo.HIV.gov, https://aidsinfo.nih.gov/understand ing-hiv-aids/glossary/456/microorganism.
7. Jay T. Lennon and Kenneth J. Locey, "There Are More Microbial Species on Earth Than Stars in the Galaxy," *Aeon*, September 10, 2018, https://aeon.co/ideas/there-are-more-microbial -species-on-earth-than-stars-in-the-sky.
8. "HIV/AIDS Glossary: Microorganism."
9. University of Georgia, "First-Ever Scientific Estimate of Total Bacteria on Earth Shows Far Greater Numbers Than Ever Known Before," *ScienceDaily*, August 25, 1998, www.sciencedaily.com /releases/1998/08/980825080732.htm.
10. Stephanie Watson, "Good vs. Bad Germs," Healthline, February 8, 2017, https://www.healthline .com/health/cold-flu/good-bad-germs.
11. Ignacio López-Goñi, "Human Microbiome: A Universe Within Us," Spanish Society for Biochemis-try and Molecular Biology, https://www.sebbm.es/revista/articulo.php?id=500&url=microbioma -humano-un-universo-en-nuestro-interior.
12. Dory Gascueña, "Microbiota: A Microscopic Shield Against COVID-19," OpenMind BBVA, April 7, 2020, https://www.bbvaopenmind.com/en/science/research/microbiota-a-microscopic -shield-against-covid-19/.
13. Lynn Margulis, "Endosymbiosis," Understanding Evolution, http://evolution.berkeley.edu /evolibrary/article/history_24.
14. Joshua Lederberg, " 'Ome Sweet 'Omics—A Genealogical Treasury of Words," *Scientist*, April 1, 2001, https://www.the-scientist.com/commentary/ome-sweet-omics--a-genealogical-treasury -of-words-54889.
15. Gabriele Berg, et al., "Microbiome Definition Revisited: Old Concepts and New Challenges," *Microbiome* 8 (June 2020), doi.org/10.1186/s40168-020-00875-0.
16. Susan E. Erdman, "Microbes, Oxytocin, and Healthful Longevity," *Journal of Probiotics & Health* 2, no. 1 (August 7, 2014), doi: 10.4172/2329-8901.1000117.
17. "Prokaryote/Procariote," Scitable, 2014, https://www.nature.com/scitable/definition/prokaryote -procariote-18/.
18. University of Georgia, "First-Ever Scientific Estimate of Total Bacteria on Earth."
19. "Antimicrobial Resistance," World Health Organization, October 13, 2020, https://www.who.int /news-room/fact-sheets/detail/antimicrobial-resistance.
20. Yan Tian, Yan Zhou, Sisi Huang, Jun Li, Kui Zhao, Xiaohui Li, Xiangchen Wen, and Xiao-an Li, "Fecal Microbiota Transplantation for Ulcerative Colitis: A Prospective Clinical Study," *BMC Gastroenterology* 19 (July 4, 2019), https://doi.org/10.1186/s12876-019-1010-4.
21. "Gastroenterology and Hepatology: Fecal Transplantation (Bacteriotherapy)," Johns Hopkins Medicine, https://www.hopkinsmedicine.org/gastroenterology_hepatology/clinical_services /advanced_endoscopy/fecal_transplantation.html.
22. Shayla Love, "Your Body Odor Might Change in Quarantine," *Vice*, April 30, 2020, https://www .vice.com/en_us/article/xgqeva/your-body-odor-might-change-in-coronavirus-quarantine.
23. "A New Naming System Proposed for Bacteria and Archaea," University of Massachusetts Am-herst News & Media Relations, June 19, 2020, https://www.umass.edu/newsoffice/article /new-naming-system-proposed-bacteria-and.

24. "Fact Sheet: The Economic Challenge Posed by Declining Pollinator Populations," White House, Office of the Press Secretary, June 20, 2014, https://obamawhitehouse.archives.gov/the-press -office/2014/06/20/fact-sheet-economic-challenge-posed-declining-pollinator-populations.
25. Nina Pullano, "Ancient Microbes in the 'Deadest' Part of Earth Redefine Boundaries of Life," *Inverse*, July 28, 2020, https://www.inverse.com/science/ancient-microbes-extreme-life-study.
26. Rafi Letzter, "Physicists: Ancient Life Might Have Escaped Earth and Journeyed to Alien Stars," *Live Science*, January 28, 2020, https://www.livescience.com/microbes-colonize-space-riding -comets.html.

01. COME TOGETHER NOW

1. History.com Editors, "Peloponnesian War," History.com, August 22, 2019, https://www.history .com/topics/ancient-history/peloponnesian-war.
2. Owen Jarus, "20 of the Worst Epidemics and Pandemics in History," *Live Science*, March 20, 2020, https://www.livescience.com/worst-epidemics-and-pandemics-in-history.html.
3. Claire Gillespie, "How Are the Spanish Flu and COVID-19 Alike? Here's What Doctors Say," *Health*, November 17, 2020, https://www.health.com/condition/infectious-diseases/how-are -spanish-flu-and-covid-19-alike.
4. Bonnie L. Bassler and Richard Losick, "Bacterially Speaking," *Cell* 125, no. 2 (April 21, 2006), https://doi.org/10.1016/j.cell.2006.04.001.
5. Wai-Leung Ng and Bonnie L. Bassler, "Bacterial Quorum-Sensing Network Architectures," *Annual Review of Genetics* 43 (2009): 197–222, doi: 10.1146/annurev-genet-102108-134304.
6. The Learning Network, "Quorum Sensing: What We Can Learn from Eavesdropping on Bacteria," *New York Times*, May 7, 2020, https://www.nytimes.com/2020/05/07/learning/quorum -sensing-what-we-can-learn-from-eavesdropping-on-bacteria.html.
7. Matthias Bauer, Johannes Knebel, Matthias Lechner, Peter Pickl, and Erwin Frey, "Ecological Feedback in Quorum-Sensing Microbial Populations Can Induce Heterogeneous Production of Autoinducers," *eLife* 6 (2017), doi: 10.7554/eLife.25773.
8. Bassler and Losick, "Bacterially Speaking."
9. Heiko Babel et al., "Ratiometric Population Sensing by a Pump-Probe Signaling System in *Bacillus subtilis*," *Nature Communications* 11 (March 4, 2020), https://doi.org/10.1038/s41467-020-14840-w.
10. Andrew M. Simons, "Modes of Response to Environmental Change and the Elusive Empirical Evidence for Bet Hedging," *Proceedings of the Royal Society B: Biological Sciences* 278, no. 1712 (March 16, 2011): 1601–9, doi: 10.1098/rspb.2011.0176.
11. Dan Cohen, "Optimizing Reproduction in a Randomly Varying Environment," *Journal of Theoretical Biology* 12, no. 1 (September 1966): 119–29, doi: 10.1016/0022-5193(66)90188-3.
12. Babel et al., "Ratiometric Population Sensing."
13. Nick Vallespir Lowery, Luke McNally, William C. Ratcliff, and Sam P. Brown, "Division of Labor, Bet Hedging, and the Evolution of Mixed Biofilm Investment Strategies," *mBio* 8, no. 4 (July/ August 2017), doi: 10.1128/mBio.00672-17.
14. Elizabeth A. Grice et al., "Topographical and Temporal Diversity of the Human Skin Microbiome," *Science* 324, no. 5931 (May 29, 2009): 1190–92, doi: 10.1126/science.1171700.
15. A. L. Cogen, V. Nizet, and R. L. Gallo, "Skin Microbiota: A Source of Disease or Defence?" *British Journal of Dermatology* 158, no. 3 (March 2008): 442–55, doi: 10.1111/j.1365-2133.2008.08437.
16. Teruaki Nakatsuji et al., "A Commensal Strain of *Staphylococcus epidermidis* Protects Against Skin Neoplasia," *Science Advances* 4, no. 2 (February 28, 2018), doi: 10.1126/sciadv.aao4502.
17. Indiana University, " 'Policing' Stops Cheaters from Dominating Groups of Cooperative Bacteria," *ScienceDaily*, May 27, 2011, https://www.sciencedaily.com/releases/2011/05/110526 103002.htm.
18. University of California-San Diego, "Synthetic Biologists Redesign the Way Bacteria 'Talk' to Each Other," Phys.org, March 5, 2020, https://phys.org/news/2020-03-synthetic-biologists -redesign-bacteria.html.
19. Ibid.
20. Liz Fuller-Wright, "Biologists Turn Eavesdropping Viruses into Bacterial Assassins," Princeton University, December 13, 2018, https://www.princeton.edu/news/2018/12/13/biologists-turn -eavesdropping-viruses-bacterial-assassins.
21. University of Tennessee at Knoxville, "Bacteria Are More Capable of Complex Decision-Making Than Thought," *ScienceDaily*, January 19, 2010, https://www.sciencedaily.com/releases /2010/01/100114143310.htm.

22. Michael Marshall, "Why Microbes Are Smarter Than You Thought," *New Scientist*, June 30, 2009, https://www.newscientist.com/article/dn17390-why-microbes-are-smarter-than-you-thought/#.

23. Inon Cohen, Ido Golding, Yonathan Kozlovsky, and Eshel Ben-Jacob, "Continuous and Discrete Models of Cooperation in Complex Bacterial Colonies," *Continuum Models and Discrete Systems: Proceedings of the 9th International Symposium*, Istanbul, Turkey, June 29-July 3, 1998, https://arxiv.org/pdf/cond-mat/9807121.pdf.

24. Denise M. Wolf, Lisa Fontaine-Bodin, Ilka Bischofs, Gavin Price, Jay Keasling, and Adam P. Arkin, "Memory in Microbes: Quantifying History-Dependent Behavior in a Bacterium," *PloS One* 3, no. 2 (2008), https://doi.org/10.1371/journal.pone.0001700.

25. University of Tennessee at Knoxville, "Bacteria Are More Capable of Complex Decision-Making Than Thought."

26. Chih-Yu Yang et al., "Encoding Membrane-Potential-Based Memory within a Microbial Community," *Cell Systems* 10, no. 5 (May 20, 2020): 417-23, doi: 10.1016/j.cels.2020.04.002.

27. Katherine Unger Baillie, "Bacteria Form Biofilms Like Settlers Form Cities," Phys.org, March 13, 2020, https://phys.org/news/2020-03-bacteria-biofilms-settlers-cities.html.

28. Trung V. Phan et al., "Bacterial Route Finding and Collective Escape in Mazes and Fractals," *Physical Review X* 10 (July 22, 2020), https://journals.aps.org/prx/abstract/10.1103/PhysRevX.10.031017.

29. Marshall, "Why Microbes Are Smarter Than You Thought."

30. Karin Sauer, "Unlocking the Secrets of Bacterial Biofilms—to Use Against Them," *Conversation*, May 31, 2016, https://theconversation.com/unlocking-the-secrets-of-bacterial-biofilms-to-use-against-them-59148.

31. José Muñoz-Dorado, Francisco J. Marcos-Torres, Elena García-Bravo, Aurelio Moraleda-Muñoz, and Juana Pérez, "Myxobacteria: Moving, Killing, Feeding, and Surviving Together," *Frontiers in Microbiology* 7 (2016), doi: 10.3389/fmicb.2016.00781.

32. Erica Bizzell, "Microbial Ninja Warriors: Bacterial Immune Evasion," American Society for Microbiology, December 2018, https://asm.org/Articles/2018/December/Microbial-Ninja-Warriors-Bacterial-Immune-Evasion.

33. Nitin Sreedhar, "When a Virus Jumps: Of Man, Microbes, and Pandemics," Livemint, April 25, 2020, https://www.livemint.com/mint-lounge/features/when-a-virus-jumps-of-man-microbes-and-pandemics-11587735889206.html.

34. Rodolphe Barrangou, "The Roles of CRISPR-Cas Systems in Adaptive Immunity and Beyond," *Current Opinion in Immunology* 32 (February 2015): 36-41, doi: 10.1016/j.coi.2014.12.008.

35. Ruairi J. Mackenzie, "DNA vs. RNA—5 Key Differences and Comparison," Technology Networks, December 18, 2020, https://www.technologynetworks.com/genomics/lists/what-are-the-key-differences-between-dna-and-rna-296719.

36. Dan Kramer, "Editing Genomes with the Bacterial Immune System," Scitable, February 9, 2013, https://www.nature.com/scitable/blog/bio2.0/editing_genomes_with_the_bacterial/.

37. Patrick D. Hsu, Eric S. Lander, and Feng Zhang, "Development and Applications of CRISPR-Cas9 for Genome Engineering," *Cell* 157, no. 6 (June 2014): 1262-78, doi: 10.1016/j.cell.2014.05.010.

38. "Bacterial Endospores," Cornell University College of Agriculture and Life Sciences, Department of Microbiology, https://micro.cornell.edu/research/epulopiscium/bacterial-endospores/.

39. R. J. Cano and M. K. Borucki, "Revival and Identification of Bacterial Spores in 25- to 40-Million-Year-Old Dominican Amber," *Science* 268, no. 5213 (May 19, 1995): 1060-64, doi: 10.1126/science.7538699.

40. "Endospore," Wikipedia, https://en.wikipedia.org/wiki/Endospore.

41. Amir Mitchell et al., "Adaptive Prediction of Environmental Changes by Microorganisms," *Nature* 460 (2009): 220-24, https://doi.org/10.1038/nature08112.

42. R. Blakemore, "Magnetotactic Bacteria," *Science* 190, no. 4212 (October 1975): 377-79, doi: 10.1126/science.170679.

43. Remy Colin, Knut Drescher, and Victor Sourjik, "Chemotactic Behaviour of *Escherichia coli* at High Cell Density," *Nature Communications* 10 (November 25, 2019), https://www.nature.com/articles/s41467-019-13179-1.

44. Klaas Bente et al., "High-Speed Motility Originates from Cooperatively Pushing and Pulling Flagella Bundles in Bilophotrichous Bacteria," *eLife* 9 (2020), https://doi.org/10.7554/eLife.47551.

45. Jennifer Tsang, "How Can a Slime Mold Solve a Maze? The Physiology Course Is Finding Out," *Well*, July 24, 2017, http://social.mbl.edu/how-can-a-slime-mold-solve-a-maze-the-physiology-course-is-finding-out.

46. Scott Chimileski and Roberto Kolter, *Life at the Edge of Sight: A Photographic Exploration of the Microbial World* (Cambridge, MA: Belknap Press, 2017).

47. Sarah D. Beagle and Steve W. Lockless, "Electrical Signalling Goes Bacterial," *Nature* 527 (October 21, 2015): 44–45, doi: 10.1038/nature15641.

48. "3.6: Some Organisms Transmit Genetic Material to Offspring without Cell Division," in Essentials of Genetics, Scitable, https://www.nature.com/scitable/ebooks/essentials-of-genetics -8/133199019/.

49. Laurence Belcher and Philip Madgwick, "Where Does Altruism Come From? Discovery of 'Greenbeard' Genes Could Hold the Answer," *Conversation*, September 11, 2019, https:// theconversation.com/where-does-altruism-come-from-discovery-of-greenbeard-genes-could -hold-the-answer-123208.

50. Jens Heller, Jiuhai Zhao, Gabriel Rosenfield, David J. Kowbel, Pierre Gladieux, and N. Louise Glass, "Characterization of Greenbeard Genes Involved in Long-Distance Kind Discrimination in a Microbial Eukaryote," *PLoS Biology* 14, no. 4 (2016), https://doi.org/10.1371/journal .pbio.1002431.

51. Ibid.

52. Derek J. Skillings, "I, Holobiont. Are You and Your Microbes a Community or a Single Entity?" *Aeon*, September 26, 2018, https://aeon.co/ideas/i-holobiont-are-you-and-your-microbes-a -community-or-a-single-entity.

53. Ibid.

54. Richard Dawkins and Yan Wong, *The Ancestor's Tale: A Pilgrimage to the Dawn of Evolution* (Boston: Mariner Books, 2016).

02. WE ARE FAMILY

1. Eric A. Franzosa, Katherine Huang, James F. Meadow, Dirk Gevers, Katherine P. Lemon, Brendan J. M. Bohannan, and Curtis Huttenhower, "Identifying Personal Microbiomes Using Metagenomic Codes," *Proceedings of the National Academy of Sciences of the United States of America*, May 11, 2015.

2. Martin J. Blaser, "Why You Should Be Worried About Changes to the Microbiome," http://www .missingmicrobes.com/.

3. "The HIVE Lab: GutFeeling KnowledgeBase," George Washington School of Medicine & Health Sciences, https://hive.biochemistry.gwu.edu/gfkb.

4. Martin J. Blaser, *Missing Microbes: How the Overuse of Antibiotics Is Fueling Our Modern Plagues* (New York: Henry Holt and Company, 2014).

5. Colleen Zacharyczuk, "Human Microbiome Project May Hold Promise for Future," Healio, May 30, 2012, https://www.healio.com/news/pediatrics/20120611/human-microbiome-project -may-hold-promise-for-future.

6. "Probiotics: Harnessing the Huge Potential of the Human Microbiome in Transforming Health and Wellness–ResearchAndMarkets.com," Business Wire, May 21, 2020, https://www.business wire.com/news/home/20200521005350/en/Probiotics-Harnessing-Huge-Potential-Human -Microbiome-Transforming.

7. Anthea Levi, "Are Synbiotics the New Probiotics?" *Health*, January 8, 2020, https://www.health .com/nutrition/synbiotics.

8. Julia Peterman, "Beyond Probiotics: Pre-, Syn-, Post-, and Psycho-," *WholeFoods Magazine*, May 22, 2019, https://wholefoodsmagazine.com/front-page/beyond-probiotics-pre-syn-post -and-psycho/.

9. Amar Sarkar, Soili M. Lehto, Siobhán Harty, Timothy G. Dinan, John F. Cryan, and Philip W. J. Burnet, "Psychobiotics and the Manipulation of Bacteria-Gut-Brain Signals," *Trends in Neurosciences* 39, no. 11 (November 2016): 763–81, doi: 10.1016/j.tins.2016.09.002.

10. P. Gualtieri et al., "Psychobiotics Regulate the Anxiety Symptoms in Carriers of Allele A of IL-1β Gene: A Randomized, Placebo-Controlled Clinical Trial," *Mediators of Inflammation* 2020, https://doi.org/10.1155/2020/2346126.

11. Jessica R. Biesiekierski, Jonna Jalanka, and Heidi M. Staudacher, "Can Gut Microbiota Composition Predict Response to Dietary Treatments?" *Nutrients* 11, no. 5 (May 2019), https://www.mdpi .com/2072-6643/11/5/1134.

12. Homepage, Viome, https://www.viome.com/.

13. Timothy G. Dinan and John F. Cryan, "Mood by Microbe: Towards Clinical Translation," *Genome Medicine* 8 (2016), doi: 10.1186/s13073-016-0292-1.

14. Shan Liang, Xiaoli Wu, and Feng Jin, "Gut-Brain Psychology: Rethinking Psychology from the Microbiota-Gut-Brain Axis," *Frontiers in Integrative Neuroscience* 12 (2018), doi: 10.3389/fnint .2018.00033.

15. Yan Wang and Lloyd H. Kasper, "The Role of Microbiome in Central Nervous System Disorders," *Brain, Behavior, and Immunity* 38 (May 2014): 1–12, doi: 10.1016/j.bbi.2013.12.015.

16. "The Brain-Gut Connection," Johns Hopkins Medicine, https://www.hopkinsmedicine.org /health/wellness-and-prevention/the-brain-gut-connection.

17. Dr. Siri Carpenter, "That Gut Feeling," American Psychological Association, September 2012, https://www.apa.org/monitor/2012/09/gut-feeling.

18. Ibid.

19. Leon G. Fine and Celine E. Riera, "Sense of Smell as the Central Driver of Pavlovian Appetite Behavior in Mammals," *Frontiers in Physiology* 10 (2019), doi: 10.3389/fphys.2019.01151.

20. "Why Is It Important to Analyze Gut Microbiota in Pregnancy?" Bioscience Institute, https:// bioinst.com/en/why-is-it-important-to-analyze-gut-microbiota-in-pregnancy/.

21. Josef Neu, MD, and Jona Rushing, MD, "Cesarean versus Vaginal Delivery: Long-Term Infant Outcomes and the Hygiene Hypothesis," *Clinics in Perinatology* 38, no. 2 (June 2011): 321–31, doi: 10.1016/j.clp.2011.03.008, https://www.ncbi.nlm.nih.gov/pmc/articles /PMC3110651/.

22. Kathy McCoy, "What Is 'Vaginal Seeding' and Will It Help Your C-Section Baby?" *Conversation*, November 28, 2019, https://theconversation.com/what-is-vaginal-seeding-and-will-it-help-your -c-section-baby-127012.

23. "The Facts About Vaginal Seeding," Grow by WebMD, February 5, 2020, https://www.webmd .com/baby/vaginal-seeding#2.

24. Ibid.

25. "The Importance of Kangaroo Care: How Skin-to-Skin Contact Is Beneficial for Infants and Caregivers," Mercy Health, August 3, 2018, https://blog.mercy.com/kangaroo-care-skin-to-skin -contact-for-infants/.

26. Priya Nimish Deo and Revati Deshmukh, "Oral Microbiome: Unveiling the Fundamentals," *Journal of Oral and Maxillofacial Pathology* 23, no. 1 (January-April 2019): 122–28, doi: 10.4103 /jomfp.JOMFP_304_18.

27. Floyd E. Dewhirst et al., "The Human Oral Microbiome," *Journal of Bacteriology* 192, no. 19 (October 2010): 5002–17, doi: 10.1128/JB.00542–10, https://www.ncbi.nlm.nih.gov/pmc /articles/PMC2944498/.

28. Bill Hesselmar, Fei Sjöberg, Robert Saalman, Nils Åberg, Ingegerd Adlerberth, and Agnes E. Wold, "Pacifier Cleaning Practices and Risk of Allergy Development," *Pediatrics* 131, no. 6 (June 2013), https://doi.org/10.1542/peds.2012-3345.

29. Li Wen and Andrew Duffy, "Factors Influencing the Gut Microbiota, Inflammation, and Type 2 Diabetes," *Journal of Nutrition* 147, no. 7 (July 2017): 1468S–75S, doi: 10.3945/jn.116.240754.

30. Kyle Bittinger et al., "Bacterial Colonization Reprograms the Neonatal Gut Metabolome," *Nature Microbiology* 5 (April 13, 2020): 838–47, https://doi.org/10.1038/s41564-020-0694-0.

31. "Obesity and Overweight: Key Facts," World Health Organization, April 1, 2020, https://www .who.int/news-room/fact-sheets/detail/obesity-and-overweight.

32. "The Microbiome and Weight Gain: Everything We Know So Far," Viome blog, March 4, 2019, https://www.viome.com/blog/microbiome-and-weight-gain-everything-we-know-so-far.

33. Darya Belikova et al., " 'Gene Accordions' Cause Genotypic and Phenotypic Heterogeneity in Clonal Populations of *Staphylococcus aureus*," *Nature Communications* 11 (2020), doi: 10.1038 /s41467-020-17277-3.

34. Ruth E. Ley et al., "Obesity Alters Gut Microbial Ecology," *Proceedings of the National Academy of Sciences of the United States of America* 102, no. 31 (August 2005): 11070–75, https://doi .org/10.1073/pnas.0504978102.

35. Rexford S. Ahima, "Digging Deeper into Obesity," *Journal of Clinical Investigation* 121, no. 6 (June 2011): 2076–79, doi: 10.1172/JCI58719.

36. Frederick W. Miller et al., "Criteria for Environmentally Associated Autoimmune Diseases," *Journal of Autoimmunity* 39, no. 4 (December 2012): 253–58, doi: 10.1016/j.jaut.2012.05.001.

37. "Autoimmune Disease," National Stem Cell Foundation, https://nationalstemcellfoundation.org /glossary/autoimmune-disease/.

38. Ruth E. Ley, "Obesity and the Human Microbiome," *Current Opinion in Gastroenterology* 26, no. 1 (January 2010): 5–11, doi: 10.1097/MOG.0b013e328333d751.

39. Alina Petre, "The Microbiome Diet: Can It Restore Your Gut Health?" Healthline, January 22, 2019, https://www.healthline.com/nutrition/microbiome-diet.
40. Sandra Dedrick et al., "The Role of Gut Microbiota and Environmental Factors in Type 1 Diabetes Pathogenesis," *Frontiers in Endocrinology* 11 (2020), doi: 10.3389/fendo.2020.00078.
41. Ibid.
42. Lenka Kramná et al., "Gut Virome Sequencing in Children with Early Islet Autoimmunity," *Diabetes Care* 38, no. 5 (May 2015): 930–33, doi: 10.2337/dc14-2490.
43. "Sardinia, Italy: Home to the World's Longest-Living Men," Blue Zones, https://www.bluezones.com/exploration/sardinia-italy/.
44. Kimberly A. Dill-McFarland et al., "Close Social Relationships Correlate with Human Gut Microbiota Composition," *Scientific Reports* 9 (2019), https://doi.org/10.1038/s41598-018-37298-9.
45. Shi Huang et al., "Human Skin, Oral, and Gut Microbiomes Predict Chronological Age," *mSystems* 5, no. 1 (January/February 2020), doi: 10.1128/mSystems.00630-19, https://msystems.asm.org/content/5/1/e00630-19.
46. Emily Mullin, "The Bacteria in Your Gut May Reveal Your True Age," *Science*, January 11, 2019, https://www.sciencemag.org/news/2019/01/bacteria-your-gut-may-reveal-your-true-age.
47. Lois Parshley, "Microbiome Science Could Bring a Revolution in Medical Care," NBCNews.com, January 26, 2017, https://www.nbcnews.com/mach/science/microbiome-science-could-bring-revolution-medical-care-n710861.
48. Martin J. Blaser, "Human Health Is in the Hands of Bacteria," *Time*, October 24, 2019, https://time.com/5709381/human-health-bacteria/.
49. Elizabeth Pennisi, "Meet the 'Psychobiome': The Gut Bacteria That May Alter How You Think, Feel, and Act," *Science*, May 7, 2020, https://www.sciencemag.org/news/2020/05/meet-psychobiome-gut-bacteria-may-alter-how-you-think-feel-and-act.
50. Ghodarz Akkasheh et al., "Clinical and Metabolic Response to Probiotic Administration in Patients with Major Depressive Disorder: A Randomized, Double-Blind, Placebo-Controlled Trial," *Nutrition* 32, no. 3 (March 2016): 315–20, http://www.sciencedirect.com/science/article/pii/S0899900715003913.
51. Michaël Messaoudi et al., "Assessment of Psychotropic-Like Properties of a Probiotic Formulation (*Lactobacillus helveticus* R0052 and *Bifidobacterium longum* R0175) in Rats and Human Subjects," *British Journal of Nutrition* 105, no. 5 (2011): 755–64, https://doi.org/10.1017/S0007114510004319.
52. Katerina V.-A. Johnson, "Gut Microbiome Composition and Diversity Are Related to Human Personality Traits," *Human Microbiome Journal* 15 (March 2020), http://www.sciencedirect.com/science/article/pii/S2452231719300181.
53. Ibid.
54. American Society for Microbiology, "Can Bacteria Make You Smarter?" *ScienceDaily*, May 25, 2010, https://www.sciencedaily.com/releases/2010/05/100524143416.htm.
55. "Microbiome-Mediated Oxytocin Release in Human Health," John Templeton Foundation, December 2019–November 2022, https://www.templeton.org/grant/microbiome-mediated-oxytocin-release-in-human-health.
56. Susan Erdman, "A Most Important Discovery," blog post, February 13, 2017, https://www.susanerdman.com/.
57. Jeffrey Norris, "Research Shows How Household Dogs Protect Against Asthma, Infection," University of California San Francisco, December 16, 2013, https://www.ucsf.edu/news/2013/12/110746/research-shows-how-household-dogs-protect-against-asthma-infection.
58. Ranjan Sinha, "What Happens to Your Microbiome If You Own a Dog?" Digbi Health, December 23, 2019, https://digbihealth.com/blogs/science-talk/what-happens-to-your-microbiome-if-you-own-a-dog.
59. Kat Eschner, "Your Home's Microbiome, Revealed," *Popular Science*, December 20, 2019, https://www.popsci.com/story/health/home-microbiome-health/.
60. Community of Microbes app, https://www.communityofmicrobes.com/app.

03. REWILDING OUR URBAN BIOME

1. Lakshmi Iyengar, "You Have a Microbial Cloud!" *Yale Scientific*, February 3, 2016, https://www.yalescientific.org/2016/02/you-have-a-microbial-cloud/.
2. James F. Meadow et al., "Humans Differ in Their Personal Microbial Cloud," *PeerJ Life & Environment* 3 (September 22, 2015), doi: 10.7717/peerj.1258.

3. Richard Conniff, "Take a Deep Breath and Say Hi to Your Exposome," *Scientific American*, September 28, 2018, https://www.scientificamerican.com/article/take-a-deep-breath-and-say-hi-to-your-exposome/.

4. Madia Lozupone et al., "The Relationship between Epigenetics and Microbiota in Neuropsychiatric Diseases," *Epigenomics* 12, no. 17 (September 9, 2020): 1559–68, https://doi.org/10.2217/epi-2020-0053.

5. Lindzi Wessel, "The Next Omics? Tracking a Lifetime of Exposures to Better Understand Disease," *Knowable Magazine*, September 19, 2019, https://www.knowablemagazine.org/article/health-disease/2019/exposome-research.

6. "Exposome and Exposomics," Centers for Disease Control and Prevention, https://www.cdc.gov/niosh/topics/exposome/default.html.

7. Conniff, "Take a Deep Breath and Say Hi to Your Exposome."

8. Chao Jiang et al., "Dynamic Human Environmental Exposome Revealed by Longitudinal Personal Monitoring," *Cell* 175 (September 20, 2018): 277–91, https://doi.org/10.1016/j.cell.2018.08.060.

9. "Architecture by Exposome," Ioana Man, https://www.ioanaman.com/architecture-by-exposome.

10. Jake M. Robinson, "Biodiversity Loss Could Be Making Us Sick—Here's Why," *Conversation*, August 4, 2020, https://theconversation.com/biodiversity-loss-could-be-making-us-sick-heres-why-143627.

11. Jacob G. Mills et al., "Relating Urban Biodiversity to Human Health with the 'Holobiont' Concept," *Frontiers in Microbiology* 10 (March 26, 2019), https://doi.org/10.3389/fmicb.2019.00550.

12. "UN Report: Nature's Dangerous Decline 'Unprecedented'; Species Extinction Rates 'Accelerating,'" United Nations, May 6, 2019, https://www.un.org/sustainabledevelopment/blog/2019/05/nature-decline-unprecedented-report/.

13. Mills et al., "Relating Urban Biodiversity to Human Health."

14. Gideon Lasco, "Could COVID-19 Permanently Change Hand Hygiene?" *Sapiens*, April 8, 2020, https://www.sapiens.org/biology/hand-hygiene-covid-19/.

15. *Report to Congress on Indoor Air Quality, Volume II: Assessment and Control of Indoor Air Pollution* (Washington, DC: US Environmental Protection Agency, August 1989).

16. "Antibacterial Soap? You Can Skip It, Use Plain Soap and Water," US Food & Drug Administration, May 16, 2019, https://www.fda.gov/consumers/consumer-updates/antibacterial-soap-you-can-skip-it-use-plain-soap-and-water#.

17. Simon Lax et al., "Longitudinal Analysis of Microbial Interaction between Humans and the Indoor Environment," *Science* 354, no. 6200 (August 29, 2014): 1048–52, doi: 10.1126/science.1254529.

18. Homepage, Map My Environment, https://www.mapmyenvironment.com/.

19. Homepage, Homebiotic, https://homebiotic.com/.

20. Katherine J. Wu, "Fight Viruses in Your Home Without Making Bacteria Stronger," *Popular Science*, April 4, 2020, https://www.popsci.com/story/diy/house-clean-bacteria-viruses-resistance/.

21. Tao Chen et al., "A Plant Genetic Network for Preventing Dysbiosis in the Phyllosphere," *Nature* 580 (2020): 653–57, doi: 10.1038/s41586-020-2185-0.

22. "Study Linking Beneficial Bacteria to Mental Health Makes Top 10 List for Brain Research," *CU Boulder Today*, January 5, 2017, https://www.colorado.edu/today/2017/01/05/study-linking-beneficial-bacteria-mental-health-makes-top-10-list-brain-research.

23. Mills et al., "Relating Urban Biodiversity to Human Health."

24. Becky Little, "When London Faced a Pandemic—And a Devastating Fire," History.com, March 25, 2020, https://www.history.com/news/plague-pandemic-great-fire.

25. *Germ City: Microbes and the Metropolis*, Museum of the City of New York, https://www.mcny.org/exhibition/germ-city.

26. Diana Budds, "How Microbes 'Designed' New York," Curbed New York, October 2, 2018, https://ny.curbed.com/2018/10/2/17925378/nyc-museum-exhibit-germs-public-housing.

27. Lucie Levine, "The Lower East Side's Forgotten Lung Block: The Italian Community Lost to 'Slum Clearance,'" 6SqFt.com, April 25, 2019, https://www.6sqft.com/the-lower-east-sides-forgotten-lung-block-the-italian-community-lost-to-slum-clearance/.

28. Roberto Kolter and E. Peter Greenberg, "The Superficial Life of Microbes," *Nature* 441 (2006): 300–302, https://www.nature.com/articles/441300a.

29. Jon Palmer, Steve Flint, and John Brooks, "Bacterial Cell Attachment, the Beginning of a Bio-

film," *Journal of Industrial Microbiology & Biotechnology* 34 (2007): 577–88, doi: 10.1007/s10295-007-0234-4.

30. Rebecca M. Landry, Dingding An, Joseph T. Hupp, Pradeep K. Singh, and Matthew R. Parsek, "Mucin-*Pseudomonas aeruginosa* Interactions Promote Biofilm Formation and Antibiotic Resistance," *Molecular Microbiology* 59, no. 1 (January 2006): 142–51, doi: 10.1111/j.1365-2958.2005.04941.x.

31. Martyn Dade-Robertson, Alona Keren-Paz, Meng Zhang, and Ilana Kolodkin-Gal, "Architects of Nature: Growing Buildings with Bacterial Biofilms," *Microbial Biotechnology* 10, no. 5 (September 2017): 1157–63, https://doi.org/10.1111/1751-7915.12833.

32. "The World's Cities in 2016," United Nations Department of Economic & Social Affairs, 2014, https://www.un.org/en/development/desa/population/publications/pdf/urbanization/the_worlds_cities_in_2016_data_booklet.pdf.

33. "Monarch Sanctuary," Terreform 1, http://terreform.org/projects_butterfly.html.

34. "Pet Industry Market Size, Trends & Ownership Statistics," American Pet Products Association (APPA), https://www.americanpetproducts.org/press_industrytrends.asp.

35. "National Feral Cat Day–October 16, 2021," National Today, https://nationaltoday.com/national-feral-cat-day/.

36. Talia Kirkland, "Cat-Astrophe! Feral Cats Amass in Philadelphia and Residents Aren't 'Amewsed,'" Fox News, May 29, 2019, https://www.foxnews.com/us/philadelphias-stray-cat-population.

37. "Pet Industry Market Size."

38. "About Pets & People," Centers for Disease Control and Prevention, https://www.cdc.gov/healthypets/health-benefits/index.html.

39. "Tokyo Metro Characteristics & Data," Metro Ad Agency, https://www.metro-ad.co.jp/en/characteristic/.

40. Chelsea Whyte, "When You Ride the Subway You Share Bacteria with Everyone in Your City," *New Scientist*, July 31, 2018, https://www.newscientist.com/article/2175556-when-you-ride-the-subway-you-share-bacteria-with-everyone-in-your-city.

41. Katherine Schulz Richard, "The Busiest Subway Systems in the World," ThoughtCo., April 10, 2019, https://www.thoughtco.com/busiest-subways-1435753.

42. "Researchers Produce First Map of New York City Subway System Microbes," Weill Cornell Medicine, February 5, 2015, https://news.weill.cornell.edu/news/2015/02/researchers-produce-first-map-of-new-york-city-subway-system-microbes-christopher-mason.

43. Isabel Reche, Gaetano D'Orta, Natalie Mladenov, Danielle M. Winget, and Curtis A. Suttle, "Deposition Rates of Viruses and Bacteria above the Atmospheric Boundary Layer," *ISME Journal* 12 (2018): 1154–62, doi: 10.1038/s41396-017-0042-4.

44. Jennifer M. Michaud et al., "Taxon-Specific Aerosolization of Bacteria and Viruses in an Experimental Ocean-Atmosphere Mesocosm," *Nature Communications* 9 (2018), https://doi.org/10.1038/s41467-018-04409-z.

45. "Life on the Wind: Study Reveals How Microbes Travel the Earth," (e)Science News, August 17, 2011, http://esciencenews.com/articles/2011/08/17/life.wind.study.reveals.how.microbes.travel.earth.

46. "Transmission of SARS-CoV-2: Implications for Infection Prevention Precautions," World Health Organization, July 9, 2020, https://www.who.int/news-room/commentaries/detail/transmission-of-sars-cov-2-implications-for-infection-prevention-precautions.

47. Predrag Slijepcevic, "Bacteria and Viruses Are Travelling the World on Highways in the Sky," *Conversation*, July 21, 2020, https://theconversation.com/bacteria-and-viruses-are-travelling-the-world-on-highways-in-the-sky-142854.

48. Ibid.

49. Jake M. Robinson, Jacob G. Mills, and Martin F. Breed, "Walking Ecosystems in Microbiome-Inspired Green Infrastructure: An Ecological Perspective on Enhancing Personal and Planetary Health," *Challenges* 9, no. 2 (2018), doi: 10.3390/challe9020040.

50. Andrew Price, "This Insane Kitchen of the Future Powers Itself with Leftovers," *Fast Company*, October 31, 2011, https://www.fastcompany.com/1678718/this-insane-kitchen-of-the-future-powers-itself-with-leftovers.

04. HERE COME THE BIOTOPIANS

1. Freeman Dyson, "Our Biotech Future," *New York Review of Books*, July 19, 2007, https://www.nybooks.com/articles/2007/07/19/our-biotech-future/.

2. Tina Hesman Saey, "CRISPR Enters Its First Human Clinical Trials," *Science News*, August 14, 2019, https://www.sciencenews.org/article/crispr-gene-editor-first-human-clinical-trials.

3. Meenakshi Prabhune, PhD, "Top CRISPR Startup Companies Changing the Future of Biotech and Medicine," Synthego, January 22, 2021, https://www.synthego.com/blog/crispr-startup-companies.

4. *Design and the Elastic Mind*, Museum of Modern Art, https://www.moma.org/calendar/exhibitions/58.

5. Homepage, Biodesign Challenge, https://biodesignchallenge.org/.

6. *Nature*, Cooper Hewitt Design Triennial with Cube Design Museum, Cooper Hewitt, https://www.cooperhewitt.org/channel/nature/.

7. "Pseudofreeze," Universidad de los Andes, Biodesign Challenge, https://biodesignchallenge.org/universidad-de-los-andes-2019.

8. "Death on Mars," University of Technology, Sydney, Biodesign Challenge, https://biodesignchallenge.org/university-of-tech-sydney.

9. Alex Pearlman, "Can We Recycle Cities?" Biodesigned, July 2020, https://www.biodesigned.org/alex-pearlman/can-we-recycle-cities.

10. "Scaling Bioprocesses," Ginkgo Bioworks, https://www.ginkgobioworks.com/our-work/scaling-bioprocesses/.

11. "UN Launches Drive to Highlight Environmental Cost of Staying Fashionable," UN News, March 25, 2019, https://news.un.org/en/story/2019/03/1035161.

12. Homepage, PILI, https://www.pili.bio/.

13. "Assemblage 002," Faber Futures, https://faberfutures.com/projects/project-coelicolor/assemblage002/.

14. "Our Revolution," Algalife, https://www.alga-life.com/our-revolution-1.

15. Elisa Allen, "The Environmental Impact of Wool," *Ecologist*, March 12, 2019, https://theecologist.org/2019/mar/12/environmental-impact-wool.

16. Western Bonime, "PETA & Stella McCartney BioDesign Challenge Winners Create Animal Free Wool," *Forbes*, July 9, 2018, https://www.forbes.com/sites/westernbonime/2018/07/09/peta-stella-mccartney-biodesign-challenge-winners-create-animal-free-wool/#4917e1d4210a.

17. "Bolt Technology—Meet B-Silk Protein," Bolt Threads, https://boltthreads.com/technology/microsilk/ and https://boltthreads.com/technology/silk-protein/.

18. Ruben Baart, "Suzanne Lee Wants to Live in a Sustainable Material World. Built with Biology, Not Oil," Next Nature Network, May 28, 2020, https://nextnature.net/story/2020/interview-suzanne-lee-2.

19. Paula Monteiro de Souza and Pérola de Oliveira Magalhães, "Application of Microbial α-Amylase in Industry—A Review," *Brazilian Journal of Microbiology* 41, no. 4 (October 2010): 850–61, doi: 10.1590/S1517-83822010000400004.

20. "Microbial Leather," Materiability Research Group, http://materiability.com/portfolio/microbial-leather/#:~:text=Thi%20biological%20alternative%20to%20leather,form%20during%20the%20drying%20process.

21. "Bolt Technology—Meet Mylo," Bolt Threads, https://boltthreads.com/technology/mylo/.

22. Jennifer Hahn, "Tōmtex Is a Leather Alternative Made from Waste Seafood Shells and Coffee Grounds," Dezeen, August 22, 2020, https://www.dezeen.com/2020/08/22/tomtex-leather-alternative-biomaterial-seafood-shells-coffee/.

23. Homepage, AlgiKnit, https://www.algiknit.com/.

24. Lilly Smith, "Forget Sewing Your Own Mask. Now You Can Grow One from Bacteria," *Fast Company*, May 16, 2020, https://www.fastcompany.com/90504945/forget-sewing-your-own-mask-now-you-can-grow-one-out-of-bacteria.

25. Rachael Rettner, "Can You Really Make 'Meat' Out of Air?" *Live Science*, November 18, 2019, https://www.livescience.com/air-protein-meat.html.

26. Augusta Pownall, "Solein Protein Powder '100 Times More Climate-Friendly' Than Other Food," Dezeen, July 3, 2019, https://www.dezeen.com/2019/07/03/solein-solar-foods-design/.

27. Ainara Sistiaga, Carolina Mallol, Bertila Galván, and Roger Everett Summons, "The Neanderthal Meal: A New Perspective Using Faecal Biomarkers," *PLoS One* 9, no. 6 (June 25, 2014), https://doi.org/10.1371/journal.pone.0101045.

28. Kelly Clime, "Beyond Sauerkraut: A Brief History of Fermented Foods," Living History Farms, March 2014, https://www.lhf.org/2014/03/beyond-sauerkraut-a-brief-history-of-fermented-foods/.

Notes ✕ 245

29. Yusuf Khan, "UBS Predicts Plant-Based Meat Sales Could Grow by More Than 25% a Year to $85 Billion by 2030," *Business Insider*, July 19, 2019, https://markets.businessinsider.com/news/stocks/beyond-meat-ubs-plant-based-meat-market-85-billion-2030-2019-7-1028367962.

30. Sindhu Raveendran et al., "Applications of Microbial Enzymes in Food Industry," *Food Technology and Biotechnology* 56, no. 1 (March 2018): 16-30, doi: 10.17113/ftb.56.01.18.5491.

31. Olivier Garret, "Missed Out on Beyond Meat? Buy These Two IPOs This Week," *Forbes*, September 11, 2019, https://www.forbes.com/sites/oliviergarret/2019/09/11/missed-out-on-beyond-meat-buy-these-2-ipos-this-week/?sh=3a6838101b60.

32. David Burrows, "Why the Food Industry Can't Ignore the Bioeconomy Boom," FoodNavigator.com, February 20, 2019, https://www.foodnavigator.com/Article/2019/02/20/Why-the-food-industry-can-t-ignore-the-bioeconomy-boom.

33. "About the Program," IndieBio, https://indiebio.co/about/.

34. Monica Watrous, "Trend of the Year: Plant-Based Foods," Food Business News, December 27, 2019, https://www.foodbusinessnews.net/articles/15105-trend-of-the-year-plant-based-foods.

35. Laura Reiley, "Impossible Burger: Here's What's Really in It," *Washington Post*, October 23, 2019, https://www.washingtonpost.com/business/2019/10/23/an-impossible-burger-dissected/.

36. Process page, Perfect Day Foods, https://www.perfectdayfoods.com/process/.

37. Kristen Catalano, "Breaking the Mold: How This Vegan Dog Food Company Is Taking on the Big Brands," This Dog's Life, October 3, 2019, https://www.thisdogslife.co/breaking-the-mold-how-this-vegan-dog-food-company-is-taking-on-the-big-brands/.

38. Alison Hewitt, "The Truth About Cats' and Dogs' Environmental Impact," UCLA Newsroom, August 2, 2017, https://newsroom.ucla.edu/releases/the-truth-about-cats-and-dogs-environmental-impact.

39. Companies page, IndieBio, https://indiebio.co/companies/.

40. Muhammad Imran Khan, Jin Hyuk Shin, and Jong Deog Kim, "The Promising Future of Microalgae: Current Status, Challenges, and Optimization of a Sustainable and Renewable Industry for Biofuels, Feed, and Other Products," *Microbial Cell Factories* 17 (2018), https://doi.org/10.1186/s12934-018-0879-x.

41. Homepage, NovoNutrients, https://www.novonutrients.com/.

42. Jenny Eagle, "Microalgae Protein Grown in Tanks to Be the Next Generation Future of Food," FoodNavigator.com, February 13, 2019, https://www.foodnavigator.com/Article/2019/02/13/Microalgae-protein-grown-in-tanks-to-be-the-next-generation-future-of-food.

43. Ibid.

44. Cheri Bantilan, "What's the Difference Between Chlorella and Spirulina?" Healthline, July 15, 2020, https://www.healthline.com/nutrition/chlorella-spirulina.

45. Homepage, FUL Foods, https://fulfoods.com.

46. Maura Judkis, "An Ocean-Obsessed Spanish Chef Brings Plankton to the Plate—and Makes It Glow," *Washington Post*, November 4, 2016, https://www.washingtonpost.com/news/food/wp/2016/11/04/an-ocean-obsessed-spanish-chef-brings-plankton-to-the-plate-and-makes-it-glow/.

47. Chelsea M. Heveran et al., "Biomineralization and Successive Regeneration of Engineered Living Building Materials," *Matter* 2, no. 2 (February 5, 2020): 481-94, doi: 10.1016/j.matt.2019.11.016.

48. Wil Srubar, "Buildings Grown by Bacteria—New Research Is Finding Ways to Turn Cells into Mini-Factories for Materials," *Conversation*, March 23, 2020, https://theconversation.com/buildings-grown-by-bacteria-new-research-is-finding-ways-to-turn-cells-into-mini-factories-for-materials-131279.

49. Homepage, Biomason, https://www.biomason.com/.

50. James Taylor-Foster, "Bricks Grown from Bacteria," *ArchDaily*, February 1, 2014, https://www.archdaily.com/472905/bricks-grown-from-bacteria.

51. Villads Egede Johansen el al., "Genetic Manipulation of Structural Color in Bacterial Colonies," *Proceedings of the National Academy of Sciences of the United States of America* 115, no. 11 (March 13, 2018): 2652-57, doi: 10,1073/pnas.1716214115.

52. Lizzie Crook, "Imperfect Perfection Lighting Range Is Made Using Bacteria," Dezeen, April 28, 2020, https://www.dezeen.com/2020/04/28/imperfect-perfection-studio-lionne-van-deursen-ventura-projects-vdf/.

53. "Three Cheers (and Three Books) for Bacteria!" Big Picture Science, September 10, 2016, https://big-picture-science.myshopify.com/blogs/news/three-cheers-and-three-books-for-bacteria.

54. Patrick D'Arcy, "Gallery: The Most Beautiful Bacteria You'll Ever See," Ideas.TED.com, March 31, 2017, https://ideas.ted.com/gallery-the-most-beautiful-bacteria-youll-ever-see/.
55. Homepage, Amino Labs, https://amino.bio/.
56. Homepage, Community of Microbes, https://www.communityofmicrobes.com/.
57. John Cumbers, "New McKinsey Report Sees a $4 Trillion Gold Rush in This One Hot Sector. Who's Selling Picks and Shovels?" *Forbes*, May 30, 2020, https://www.forbes.com/sites/john cumbers/2020/05/30/mckinsey-report-4-trillion-gold-rush-bioeconomy-synthetic-biology /?sh=56e05d7d4cfd.
58. Homepage, BioCurious, https://biocurious.org/.
59. Homepage, Genspace, https://www.genspace.org/.
60. Homepage, iGEM, https://igem.org/Main_Page.
61. Homepage, OpenCell, https://www.opencell.bio/.
62. Westley Dang, "AsimicA: Raising the Bar for All Biofermentation," IndieBio blog, October 28, 2020, https://indiebio.co/asimica-raising-the-bar-for-all-biofermentation/.
63. Homepage, AsimicA, https://asimica.com/index.html.

05. ANCIENT HIGH TECHNOLOGY

1. Colin Barras, "Bacteria: The Miracle Microbes That Could Fix Planet," *Science Focus*, September 2, 2020, https://www.sciencefocus.com/science/bacteria-the-miracle-microbes-that-could -fix-planet.
2. Anand Jagatia, "Archaea and the Tree of Life," Microbiology Society, May 9, 2016, https://micro biologysociety.org/blog/archaea-and-the-tree-of-life.html.
3. Camille Barr, *Thermus aquaticus*," Montana Natural History Center, February 25, 2018, https:// www.montananaturalist.org/blog-post/thermus-aquaticus/.
4. Ibid.
5. Jeff Havig and Trinity Hamilton, "How a Thermophilic Bacterium from a Yellowstone Hot Spring Is Helping the Fight Against the COVID-19 Pandemic," US Geological Survey (USGS), March 30, 2020, https://www.usgs.gov/center-news/how-a-thermophilic-bacterium-a-yellowstone-hot -spring-helping-fight-against-covid-19.
6. Ivy Engel, "How a Bacteria from Yellowstone Is Helping Fight Coronavirus," Wyoming Public Media, April 20, 2020, https://www.wyomingpublicmedia.org/post/how-bacteria-yellowstone -helping-fight-coronavirus#stream/0.
7. "Nobel-Winning Research in the Natural Laboratory That Is Yellowstone," US Geological Survey (USGS), May 7, 2018, https://www.usgs.gov/center-news/nobel-winning-research-natural -laboratory-Yellowstone.
8. "Biomass Explained," US Energy Information Administration, https://www.eia.gov/energy explained/biomass/.
9. Cedric Jasper Hahn et al., " 'Candidatus Ethanoperedens,' a Thermophilic Genus of *Archaea* Mediating the Anaerobic Oxidation of Ethane," *mBio* 11, no. 2 (March/April 2020), doi: 10.1128 /mBio.00600-20, https://www.mpg.de/14670461/0409-mbio-064278-new-ethane-munching -microbes-discovered-at-hot-vents.
10. "Swiss Researchers Identify New Bacteria in Permafrost," SWI SwissInfo.ch, June 20, 2020, https://www.swissinfo.ch/eng/swiss-researchers-identify-new-bacteria-in-permafrost /45849866.
11. Claire A. Nordeen and Sandra L. Martin, "Engineering Human Stasis for Long-Duration Spaceflight," *Physiology* 34, no. 2 (March 2019): 101–11, https://journals.physiology.org/doi /abs/10.1152/physiol.00046.2018.
12. Shlee S. Song and Patrick D. Lyden, "Overview of Therapeutic Hypothermia," *Current Treatment Options in Neurology* 14, no. 6 (December 2012): 541–48, doi: 10.1007/s11940-012-0201-x.
13. "Out in the Cold," Harvard Health Letter, January 2010, https://www.health.harvard.edu/staying -healthy/out-in-the-cold.
14. "Swiss Researchers Identify New Bacteria in Permafrost."
15. Ricardo Cavicchioli et al., "Scientists' Warning to Humanity: Microorganisms and Climate Change," *Nature Reviews Microbiology* 17 (2019): 569–86, doi: 10.1038/s41579-019-0222-5.
16. University of New South Wales, "Leaving Microbes Out of Climate Change Conversation Has Major Consequences, Experts Warn," *ScienceDaily*, June 18, 2019, https://www.sciencedaily .com/releases/2019/06/190618113126.htm.
17. Shmuel Gleizer et al., "Conversion of *Escherichia coli* to Generate All Biomass Carbon from

CO$_2$," *Cell* 179, no. 6 (November 27, 2019): 1255–63, https://doi.org/10.1016/j.cell.2019
.11.009.

18. Renee Cho, "Can Soil Help Combat Climate Change?" *State of the Planet*, Columbia University
Climate School, February 21, 2018, https://blogs.ei.columbia.edu/2018/02/21/can-soil-help
-combat-climate-change/#.

19. Krisy Gashler, "Newly Found Bacteria Fights Climate Change, Soil Pollutants," *Cornell Chronicle*,
February 20, 2020, https://news.cornell.edu/stories/2020/02/newly-found-bacteria-fights
-climate-change-soil-pollutants.

20. Mark Acosta, "The Problem with Bottled Water," *Environment Today Magazine*, https://www.ets
.org/s/research/pdf/reading4_bottled_water.pdf.

21. Einat Paz-Frankel, "Genetically Modified Bacteria Could Eat Away the World's Massive Plastic
Problem," NoCamels, January 22, 2017, https://nocamels.com/2017/01/genetically-modified
-bacteria-eat-plastic/.

22. Ibid.

23. María José Cárdenas Espinosa et al., "Toward Biorecycling: Isolation of a Soil Bacterium That
Grows on a Polyurethane Oligomer and Monomer," *Frontiers in Microbiology* 11 (March 2020),
https://www.frontiersin.org/article/10.3389/fmicb.2020.00404.

24. Barras, "Bacteria: The Miracle Microbes That Could Fix Planet."

25. Leigh Krietsch Boerner, "BioCollection's Miranda Wang and Jeanny Yao Aim to Make Treasure
Out of Plastic Trash," *Chemical & Engineering News* (*C&EN*), March 8, 2020, https://cen.acs.org
/environment/recycling/BioCollection-Miranda-Wang-and-Jeanny-Yao-aim-to-make-treasure
-out-of-plastic-trash/98/i9.

26. Laura Parker, "A Whopping 91% of Plastic Isn't Recycled," *National Geographic*, December 20,
2018, https://www.nationalgeographic.com/news/2017/07/plastic-produced-recycling-waste
-ocean-trash-debris-environment/.

27. Homepage, Scindo, https://scindo.bio/.

28. "The Role of Bacteria in a Healthy Septic System," West Coast Sanitation Inc., March 1, 2018,
https://westcoastsanitationinc.com/the-role-of-bacteria-in-a-healthy-septic-system/.

29. Barras, "Bacteria: The Miracle Microbes That Could Fix Planet."

30. Gege Li, "Soil Gets Its Smell from Bacteria Trying to Attract Invertebrates," *New Scientist*, April
6, 2020, https://www.newscientist.com/article/2239854-soil-gets-its-smell-from-bacteria-trying
-to-attract-invertebrates/.

31. Stephen C. Wagner, "Biological Nitrogen Fixation," *Nature Education Knowledge Project* 3,
no. 10 (2011): 15, https://www.nature.com/scitable/knowledge/library/biological-nitrogen
-fixation-23570419/.

32. Laura Bergshoef, "Engineering Soil Bacteria Could Help Develop Enhanced, 'Non-GMO' Crops
for Europe," *New Scientist*, October 26, 2020, https://geneticliteracyproject.org/2020/10/26
/engineering-soil-bacteria-could-help-develop-enhanced-non-gmo-crops-for-europe/.

33. Homepage, Reazent, http://reazent.com/.

34. Homepage, Allied Microbiota, https://www.alliedmicrobiota.com/.

35. Christopher McFadden, "7 Important Biofuel Crops That We Use for Fuel Production," Interesting Engineering, April 2, 2021, https://interestingengineering.com/seven-biofuel-crops-use
-fuel-production.

36. Brajesh Singh, Fernando T. Maestre, and Manuel Delgado Baquerizo, "If the World's Soils
Keep Drying Out That's Bad News for Microbes (and People)," *Conversation*, February 8, 2016,
https://theconversation.com/if-the-worlds-soils-keep-drying-out-thats-bad-news-for-microbes
-and-people-53937.

37. "Renewable Energy Explained," US Energy Information Administration, https://www.eia.gov
/energyexplained/renewable-sources/.

38. Katie Gerhards, "Building Better Bacteria for Bioenergy Production," University of Wisconsin-
Madison School of Pharmacy, March 7, 2019, https://pharmacy.wisc.edu/building-better
-bacteria-for-bioenergy-production/.

39. Ibid.

40. Abdelrhman Mohamed, Phuc T. Ha, Brent M. Peyton, Rebecca Mueller, Michelle Meagher, and
Haluk Beyenal, "*In situ* Enrichment of Microbial Communities on Polarized Electrodes Deployed
in Alkaline Hot Springs," *Journal of Power Sources* 414 (February 28, 2019): 547–56, http://www
.sciencedirect.com/science/article/pii/S0378775319300291.

41. Khan, Shin, and Kim, "The Promising Future of Microalgae."

42. Xiaomeng Liu et al., "Power Generation from Ambient Humidity Using Protein Nanowires," *Nature* 578 (2020): 550-54, https://doi.org/10.1038/s41586-020-2010-9.
43. Sophia Ktori, "Bio-Batteries: Creating Energy from Bacteria," Engineering & Technology, July 15, 2013, https://eandt.theiet.org/content/articles/2013/07/bio-batteries-creating-energy-from-bacteria/.
44. Nikhil S. Malvankar and Derek R. Lovley, "Microbial Nanowires: A New Paradigm for Biological Electron Transfer and Bioelectronics," *ChemSusChem* 5, no. 6 (June 2012): 1039-46, https://www.micro.umass.edu/faculty-and-research/publications/microbial-nanowires-a-new-paradigm-for-biological-electron.
45. "What If Microbes Were Smarter Than We Thought?" University of Toronto Temerty Faculty of Medicine, https://medicine.utoronto.ca/about-faculty-medicine/what-if-microbes-were-smarter-we-thought.
46. Liping Zhu, Song-Ju Kim, Masahiko Hara, and Masashi Aono, "Remarkable Problem-Solving Ability of Unicellular Amoeboid Organism and Its Mechanism," *Royal Society Open Science* 5, no. 12 (December 2018), https://royalsocietypublishing.org/doi/10.1098/rsos.180396.
47. "Antimicrobial Resistance," World Health Organization, October 13, 2020, https://www.who.int/news-room/fact-sheets/detail/antimicrobial-resistance.
48. Anne Trafton, "Artificial Intelligence Yields New Antibiotic," *MIT News*, February 20, 2020, https://news.mit.edu/2020/artificial-intelligence-identifies-new-antibiotic-0220.
49. Raphael Kim and Stefan Poslad, "The Thing with *E. coli*: Highlighting Opportunities and Challenges of Integrating Bacteria in IoT and HCI," Cornell University, June 17, 2019, https://arxiv.org/abs/1910.01974.
50. Robert Sanders, "$275 Million Commitment to Brew Better Molecules for Manufacturing," *Berkeley News*, October 20, 2020, https://news.berkeley.edu/2020/10/20/275-million-commitment-to-brew-better-molecules-for-manufacturing/.
51. "DOD Approves $87 Million for Newest Bioindustrial Manufacturing Innovation Institute," US Department of Defense, October 20, 2020, https://www.defense.gov/Newsroom/Releases/Release/Article/2388087/dod-approves-87-million-for-newest-bioindustrial-manufacturing-innovation-insti/.
52. Shannon Ellis, "Biotech Booms in China," *Nature* 553, no. 7688 (January 17, 2018): S19-S22, https://www.nature.com/articles/ d41586-018-00542-3.
53. "India—a Biotech Growth Catalyst," Invest India, https://www.investindia.gov.in/sector/bio technology#.
54. Frédéric Simon, "EU to Unveil Trillion-Euro 'Green Deal' Financial Plan," Euractiv.com, January 14, 2020, https://www.euractiv.com/section/energy-environment/news/eu-to-unveil-trillion-euro-green-deal-financial-plan/.

06. MICRO-NAUTS

1. Addy Pross and Robert Pascal, "The Origin of Life: What We Know, What We Can Know and What We Will Never Know," *Open Biology* 3, no. 3 (March 2013), doi: 10.1098/rsob.120190.
2. " 'Oumuamua," NASA Science Solar System Exploration, December 19, 2019, https://solar system.nasa.gov/asteroids-comets-and-meteors/comets/oumuamua/in-depth/.
3. "In Search of Panspermia," Astrobiology at NASA, April 16, 2021, https://astrobiology.nasa.gov/news/in-search-of-panspermia/.
4. Stephen Fleischfresser, "A Brief History of Panspermia," *Cosmos*, April 23, 2018, https://cosmos magazine.com/biology/over-our-heads-a-brief-history-of-panspermia/.
5. Ibid.
6. "Important Scientists: Fred Hoyle (1915-2001)," Physics of the Universe, https://www.physic softheuniverse.com/scientists_hoyle.html.
7. Brig Klyce, "Bacteria: The Space Colonists," Panspermia.org, https://www.panspermia.org/bacteria.htm.
8. "Surveyor Crater and Surveyor III," Apollo 12 Lunar Surface Journal, http://www.hq.nasa.gov/office/pao/History/alsj/a12/a12.surveyor.html.
9. Leonard David, "Moon Microbe Mystery Finally Solved," Space.com, May 2, 2011, http://www.space.com/11536-moon-microbe-mystery-solved-apollo-12.html.
10. F. J. Mitchell and W. L. Ellis, "Surveyor III: Bacterium Isolated from Lunar-Retrieved TV Camera," *Proceedings of the Second Lunar Science Conference* 3 (1971): 2721-33, http://adsabs.harvard.edu/full/1971LPSC. . . . 2.2721M.

11. Chandra Wickramasinghe, PhD; Kamala Wickramasinghe, MA; and Gensuke Tokoro, *Our Cosmic Ancestry in the Stars: The Panspermia Revolution and the Origins of Humanity* (Rochester, VT: Bear & Company, 2019).

12. Carrie Gilder, "Study Finds Space Station Microbes Are No More Harmful Than Those Found in Similar Ground Environments," NASA, January 22, 2020, https://www.nasa.gov/mission _pages/station/research/news/space-station-microbes-no-more-harmful-than-those-on-earth -extremophiles.

13. Aram Avila-Herrera et al., "Crewmember Microbiome May Influence Microbial Composition of ISS Habitable Surfaces," *PLoS One* 15, no. 4 (2020), https://doi.org/10.1371/journal.pone .0231838.

14. Alexandra Witze, Smriti Mallapaty, and Elizabeth Gibney, "Countdown to Mars: Three Daring Missions Take Aim at the Red Planet," *Nature*, July 7, 2020, https://www.nature.com/articles /d41586-020-01861-0.

15. Hanneke Weitering, "Elon Musk Says SpaceX's 1st Starship Trip to Mars Could Fly in 4 Years," Space.com, October 17, 2020, https://www.space.com/spacex-starship-first-mars-trip-2024.

16. Paul Chambers, *Life on Mars: The Complete Story* (London: Blandford Press, 1999).

17. Gilbert V. Levin and Patricia Ann Straat, "Viking Labeled Release Biology Experiment: Interim Results," *Science* 194, no. 4271 (December 17, 1976): 1322-29, doi: 10.1126/science.194 .4271.1322, https://science.sciencemag.org/content/194/4271/1322.

18. Gilbert V. Levin and Patricia Ann Straat, "Completion of the Viking Labeled Release Experiment on Mars," *Journal of Molecular Evolution* 14, no. 1 (March 1979): 167–83, doi: 10.1007/BF017 32376, https://link.springer.com/article/10.1007%2FBF01732376.

19. Richard Stenger, "Mars Sample Return Plan Carries Microbial Risk, Group Warns," CNN, November 7, 2000, http://edition.cnn.com/2000/TECH/space/11/07/mars.sample/.

20. Ibid.

21. Gilbert V. Levin, "I'm Convinced We Found Evidence of Life on Mars in the 1970s," *Scientific American*, October 10, 2019, https://blogs.scientificamerican.com/observations/im-convinced -we-found-evidence-of-life-on-mars-in-the-1970s/.

22. Aylin Woodward, "Before We Put People on Mars, We Should Infect the Planet with Earthly Microbes, a Group of Scientists Says," *Business Insider*, October 2, 2019, https://www.business insider.com/mars-infected-with-microbes-before-people-land-2019-10.

23. Jose V. Lopez, Raquel S. Peixoto, and Alexandre S. Rosado, "Inevitable Future: Space Colonization Beyond Earth with Microbes First," *FEMS Microbiology Ecology* 95, no. 10 (October 2019), https://doi.org/10.1093/femsec/fiz127.

24. Sarah Bordenstein, "Tardigrades (Water Bears)," Microbial Life Educational Resources, http:// serc.carleton.edu/microbelife/topics/tardigrade/index.html.

25. Daniel Oberhaus, "A Crashed Israeli Lunar Lander Spilled Tardigrades on the Moon," *Wired*, August 5, 2019, https://www.wired.com/story/a-crashed-israeli-lunar-lander-spilled-tardigrades -on-the-moon/?verso=true.

26. Rachel Courtland, " 'Water Bears' Are First Animal to Survive Space Vacuum," *New Scientist*, September 8, 2008, https://www.newscientist.com/article/dn14690-water-bears-are-first-animal -to-survive-space-vacuum/#ixzz6eprzFKCu.

27. Sebastian Emanuel Lauro et al., "Multiple Subglacial Water Bodies Below the South Pole of Mars Unveiled by New MARSIS Data," *Nature Astronomy* 5 (September 28, 2020): 63-70, https://doi.org/10.1038/s41550-020-1200-6.

28. "Mars Facts," NASA Science Mars Exploration Program, https://mars.nasa.gov/all-about-mars /facts/.

29. E. Heydari et al., "Deposits from Giant Floods in Gale Crater and Their Implications for the Climate of Early Mars," *Scientific Reports* 10 (November 2020), https://doi.org/10.1038/s41598 -020-75665-7.

30. J. Benton Jones Jr., *Hydroponics: A Practical Guide for the Soilless Grower*, 2nd ed. (Boca Raton, FL: CRC Press, 2005), 153-66.

31. Robert Z. Pearlman, "Astronauts Take First Bites of Lettuce Grown in Space," *Scientific American*, August 10, 2015, https://www.scientificamerican.com/article/astronauts-take-first-bites-of -lettuce-grown-in-space/#.

32. "NASA's SpaceX Crew-1 Astronauts Headed to International Space Station," NASA, November 15, 2020, https://www.nasa.gov/press-release/nasa-s-spacex-crew-1-astronauts-headed -to-international-space-station/.

33. Jeremy Rehm, "Radar Points to Moon Being More Metallic Than Researchers Thought," NASA, July 1, 2020, https://www.nasa.gov/feature/goddard/2020/moon-more-metallic-than-thought.
34. Charles S. Cockell et al., "Space Station Biomining Experiment Demonstrates Rare Earth Element Extraction in Microgravity and Mars Gravity," *Nature Communications* 11 (2020), https://doi.org/10.1038/s41467-020-19276-w.
35. "ArianeGroup to Study a Moon Mission for ESA," ArianeGroup, January 21, 2019, https://www.ariane.group/en/news/arianegroup-to-study-a-moon-mission-for-esa/.
36. "China to Launch Chang'e 5 Lunar Probe This Year," XinhuaNet, September 19, 2020, http://www.xinhuanet.com/english/2020-09/19/c_139381321.htm.
37. Cockell et al., "Space Station Biomining Experiment."
38. Robert Sanders, "On Mars or Earth, Biohybrid Can Turn CO_2 into New Products," *Berkeley News*, March 31, 2020, https://news.berkeley.edu/2020/03/31/on-mars-or-earth-biohybrid-can-turn-co2-into-new-products.
39. Lonnie Shekhtman, "NASA Scientists Confirm Water Vapor on Europa," NASA, November 18, 2019, https://www.nasa.gov/feature/goddard/2019/nasa-scientists-confirm-water-vapor-on-europa.
40. "Venus," NASA Science Solar System Exploration, December 19, 2019, https://solarsystem.nasa.gov/planets/venus/in-depth/.
41. Jane S. Greaves et al., "Phosphine Gas in the Cloud Decks of Venus," *Nature Astronomy* (September 2020), https://doi.org/10.1038/s41550-020-1174-4.
42. "In Memoriam: Harold Morowitz," Santa Fe Institute, March 30, 2016, https://www.santafe.edu/news-center/news/memoriam-harold-morowitz.
43. Terry Devitt, "Is There Life Adrift in the Clouds of Venus?" University of Wisconsin–Madison News, March 30, 2018, https://news.wisc.edu/is-there-life-adrift-in-the-clouds-of-venus/.
44. Elissa D. Pastuzyn et al., "The Neuronal Gene *Arc* Encodes a Repurposed Retrotransposon Gag Protein That Mediates Intercellular RNA Transfer," *Cell* 172, nos. 1–2 (January 11, 2018): 275–88, https://www.cell.com/cell/fulltext/S0092-8674(17)31504-0.
45. Stephanie Pappas, "Unraveling the Human Genome: 6 Molecular Milestones," *Live Science*, January 23, 2013, https://www.livescience.com/26505-human-genome-milestones.html.
46. Pastuzyn et al., "The Neuronal Gene *Arc*."
47. A. Berliner, T. Mochizuki, and K. Stedman, "Astrovirology: Viruses at Large in the Universe," *Astrobiology* 18, no. 2 (February 2018): 207–23, https://astrobiology.nasa.gov/nai/media/medialibrary/2019/10/StedmanAstrovirologyIntro.pdf.

07. ALICE IN FUTURELAND

1. Fergus Walsh, "Superbugs to Kill 'More Than Cancer' by 2050," BBC News, December 11, 2014, https://www.bbc.com/news/health-30416844.

Illustration Credits

Most of the photographs throughout this book were sourced from Pexels.com, a free stock photo and video website. Thank you to the photographers and individuals who support open-source photography organizations.

00. INTRODUCTION

pp. viii–ix: Welcome on board. Photo by Mabel Amber, from Pexels; Human hands and dog paw. Photo by Anastasiya Lobanovskaya, from Pexels; Man with cup to ear. Photo by Andrea Piacquadio, from Pexels; Spoon with colored sprinkles. Photo by Isabella Clifford, from Pexels; Yellow spores on blue mass. Photo from Unsplash; Seated person looking at petri dish on iPad. Photo by Retha Ferguson, from Pexels.

p. x: Blue hand/red background. Photo by Cottonbro, from Pexels.

p. xi: Sign of "One World" and picture of Earth. Photo by Markus Spiske, from Pexels.

p. xii: Aerial view of swimmers. Photo by Kelly Lacy, from Pexels.

p. xiv: Blue book on pink background. Photo by Ann H, from Pexels.

p. xv: Night sky with stars. Photo by Juan, from Pexels.

p. xvi: Man at table reaching for red comment bubble. Photo by Oladimeji Ajegbile, from Pexels.

p. xviii: Close-up of human skin with hair. Photo by Karolina Grabowska, from Pexels.

p. xx: Close-up of eye. Photo by Maksin Goncharenok, from Pexels.

p. xxii: Orange flag waving in sky. Photo by Karolina Grabowska, from Pexels.

p. xxiii: Aerial view of city at night. Photo by Eberhard Grossgasteiger, from Pexels.

p. xxv: Magnifying glass in front of starry sky. Photo by Rakicevic Nenad, from Pexels.

p. xxvii: Hand with glass jar releasing light into starry sky. Photo by Rakicevic Nenad, from Pexels.

p. xxviii: Welcome on board. Photo by Mabel Amber, from Pexels.

01. COME TOGETHER NOW

p. 2: Orange sphere. Photo by Benjamin Suter, from Pexels.

p. 3: Silver pendulum balls. Photo by Pixabay, from Pexels.

p. 5: Out-of-focus lights against black background. Photo by Madison Inouye, from Pexels.

p. 7: Green crosswalk sign. Photo by Jeshoots.com, from Pexels.

p. 8: Side view of man's face. Photo by Ksenia Chernaya, from Pexels.

p. 10: Blue bacteria. Photo by cdc-QEU-QglOJKA, from Unsplash.

p. 12: Neon laptop against black background. Photo by Junior Teixeira, from Pexels.

p. 13: Theater mask in grass. Photo by Magda Ehlers, from Pexels.

p. 14: Boy drawing bacteria on whiteboard. Photo by Katerina Holmes, from Pexels.

p. 17: Child in red coat walking through forest. Photo by Ksenia Chernaya, from Pexels.

p. 18: Orange-handled tools on peg wall. Photo by Polesie Toys, from Pexels.

p. 22: Dark cloud against orange and blue sunset. Photo by Disha Sheta, from Pexels.

p. 24: Scissors and spools of thread. Photo by Gustavo Fring, from Pexels.

p. 26: Bronze powder swirl. Photo by Karolina Grabowska, from Pexels.

pp. 28–29: Long-exposure nighttime highway. Photo by Pixabay, from Pexels.

p. 30: Two hands clasping. Photo by Ave Calvar Martinez, from Pexels.

p. 31: Bearded man with glasses with algorithm cast in green over face. Photo by Cottonbro, from Pexels.

p. 33: Mushrooms growing on rock. Photo by Valeriia Miller, from Pexels.

p. 34: Hand reflected in mirror. Photo by Cottonbro, from Pexels.

p. 36: Orange coral in blue water. Photo by N-i-e-r-o-s-h-o-t-s, from Pexels.

pp. 38–39: Gray wall with hand silhouetted in oval mirror. Photo by Emre Can, from Pexels.

p. 40: Human hands and pet paw. Photo by Anastasiya Lobanovskaya, from Pexels.

02. WE ARE FAMILY

p. 42: Girl with fingers on face. Photo by Ike Louie Nativda, from Pexels.
p. 44: Woman with plant on shoulder. Photo by Cottonbro, from Pexels.
p. 47: Pink slippers with photos on floor. Photo by Lisa Fotios, from Pexels.
p. 48: Cracked dry earth with green leaves growing. Photo by Iconcom, from Pexels.
p. 49: Girl's dirty hands holding potted plant in front of face. Photo by Valeria Ushakova, from Pexels.
p. 50: Aerial shot of various foods in bowls. Photo by Sewdream, from Adobe.
p. 52: Hand tossing roll of toilet paper. Photo by Anna Svhets, from Pexels.
p. 54: Woman stretching to sky in front of trees. Photo by Retha Ferguson, from Pexels.
pp. 56-57: Spoons holding herbs and spices. Photo by Karolina Grabowska, from Pexels.
pp. 58-59: Close-up of woman in striped sweater cradling pregnant belly. Free Stock from Pexels.
p. 60: Close-up of baby on mother's bare shoulder. Photo by RayCan, from Adobe.
p. 60: Mom and baby smiling. Photo by Edward Ever, from Pexels.
p. 61: Tan fabric with colored felt letters spelling "smile." Photo by Magda Ehlers, from Pexels.
p. 63: Couple with infant. Photo by Nappy, from Pexels.
p. 64: Bare feet on scale. Photo by Ketut Subiyanto, from Pexels.
p. 66: Couple feeding each other. Photo by Cottonbro, from Pexels.
p. 68: Map with Polaroids. Photo by Leah Kelley, from Pexels.
p. 70: Orange corkscrew birthday candle. Photo by Karolina Grabowska, from Pexels.
p. 72: Hand pulling arrows out of target. Photo by Burst, from Pexels.
p. 74: Woman in blue top touching hair. Photo by Andrew Neel, from Pexels.
p. 75: Young girl in pink dress drinking from cup. Photo by Ksenia Chernaya, from Pexels.
p. 76: Woman pointing to her temple. Photo by Engin Akyurt, from Pexels.
p. 78: People eating at sidewalk table. Photo by Daria Sannikova, from Pexels.
p. 79: Young boy riding orange bike. Photo by Tarikul Raana, from Pexels.
p. 81: Red heart candies spilling out of bottle. Photo by Dear W, from Pexels.
p. 82: Woman hugging dog, pink background. Photo by Daria Shetsova, from Pexels.
p. 84: Barefoot person with sheet over body. Photo by FOTOGRAFIERENDE, from Pexels.
p. 86: Yellow spores on blue mass. Photo from Unsplash.

03. REWILDING OUR URBAN BIOME

p. 88: Couple silhouetted against sunset. Photo by Dio Alif Utomo, from Pexels.
p. 89: Woman superimposed over road. Photo by Ash Cork, from Pexels.
p. 91: Family silhouetted against sky. Photo by Charles Pragnell, from Pexels.
p. 93: Girls tossing hoops in playground. Photo by Kamille Samaio, from Pexels.
p. 93: Girl wearing face mask. Photo by RFstudio, from Pexels.
p. 94: Clay tiles with green leaf growing in crack. Photo by sum+it, from Pexels.
p. 99: Silhouette of hand opening door. Photo by George Becker, from Pexels.
p. 100: Window view from dark room. Photo by Cottonbro, from Pexels.
p. 103: Houseplants on windowsill. Photo by Ksenia Chernaya, from Pexels.
p. 104: Woman sitting on windowsill looking at Empire State Building. Photo by Taryn Elliott, from Pexels.
p. 107: Aerial view of coastline. Photo by Tom Fisk, from Pexels.
p. 109: Gold butterfly on leaf. Photo by SK, from Pexels.
p. 111: Dog running with green ball in mouth. Photo by Chepté Cormani, from Pexels.
p. 112: Subway car interior. Photo by Renz Macorol, from Pexels.
p. 113: Woman watching train go by. Photo by Jeffrey Czum, from Pexels.
p. 115: Woman blowing colored dust. Photo by Cottonbro, from Pexels.
p. 115: Rust-colored cloudy sky. Photo by Daniel Maldonado, from Pexels.
p. 116: Pink and blue sky with plane trailing white plume. Photo by Eberhard Grossgasteiger, from Pexels.
p. 118: Looking up in atrium. Photo by mentatdgt, from Pexels.
p. 119: Woman opening white refrigerator. Photo by Cottonbro, from Pexels.
p. 120: Seated person looking at petri dish on iPad. Photo by Retha Ferguson, from Pexels.

04. HERE COME THE BIOTOPIANS

p. 122: Man in white turtleneck covered in red string. Photo by Vitoria Santos, from Pexels.
p. 125: Fuchsia spores on green stalk. Photo by Pixabay, from Pexels.

p. 125: Abstract circle art with man walking by. Photo by Ivan Rivero, from Pexels.

p. 127: Palms with "yes" and "no" in white. Photo by Cottonbro, from Pexels.

p. 128: Hands assembling plastic molecule puzzle. Photo by RFStudio, from Pexels.

p. 129: Closed gray shutter with light spots. Photo by Pixabay, from Pexels.

p. 131: Man holding green earthy textile. Photo by thisisengineering, from Pexels.

p. 133: Spiderweb. Photo by Skitterphoto, from Pexels.

p. 135: Rolled black, white, and tan fabrics on table. Photo by Cottonbro, from Pexels.

p. 136: Sliced bread. Photo by Mariana Kurnyk, from Pexels.

p. 137: Hands grabbing grains. Photo by Cottonbro, from Pexels.

p. 141: Cloud in an ice cream cone. Photo by Rakicevic Nenad, from Pexels.

p. 144: Dalmatian getting a treat. Photo by Kasuma, from Pexels.

p. 145: Green plants under water. Photo by Anni Roenkae, from Pexels.

p. 147: Aerial view of green lattes. Photo by Eunice Lui, from Pexels.

p. 149: Gray room with photo of green fern and table with laptop. Photo by Emma Pollard, from Pexels.

p. 151: Water-filled sphere on beach. Photo by Josh Sorenson, from Pexels.

pp. 152-53: Red fish with fanned tail. Photo by Chevanon Photography, from Pexels.

p. 154: Child playing with small wooden animals. Photo by Cottonbro, from Pexels.

p. 158: Hand on fogged window in front of colored lights. Photo by Josh Hild, from Pexels.

p. 160: Man with cup to ear. Photo by Andrea Piacquadio, from Pexels.

05. ANCIENT HIGH TECHNOLOGY

p. 162: Green laser. Photo by Pixabay, from Pexels.

p. 163: Orange strobes over blurred image of man. Photo by Aidan Roof, from Pexels.

p. 164: Petri dish with black specks in orange and light blue fluid. Photo by Edward Jenner, from Pexels.

p. 166: Hot springs with orange water. Photo by Pixabay, from Pexels.

p. 167: Sunlight shining through water. Photo by Jeremy Bishop, from Pexels.

pp. 168-69: Blue ice with mountains. Photo by Pixabay, from Pexels.

p. 170: Aerial view of green soccer field. Photo by Mike, from Pexels.

p. 172: Hand holding globe in blue plastic bag. Photo by Anna Shvets, from Pexels.

p. 175: Trowel scooping soil. Photo by Lisa Fotios, from Pexels.

p. 177: Person wearing blue jeans and dirty white sneakers. Photo by Ricardo Esquivel, from Pexels.

p. 178: Yellow liquid on white background. Photo by Karolina Grabowska, from Pexels.

p. 179: Light bulbs on wire. Photo by Ion Ceban, from Pexels.

p. 181: Colored wooden blocks with numbers. Photo by Digital Buggu, from Pexels.

p. 182: Spiraled lights on dark highway. Photo by Pixabay, from Pexels.

p. 185: Circular red light on circuitry. Photo by Jean-Pierre Barthe, from Adobe.

p. 185: Colorful lines of code. Photo by Marcus Spiske, from Pexels.

p. 187: View of trees through hole in white ceiling. Photo by Daria Shevtsova, from Pexels.

p. 189: Hand pouring seeds out of cup. Photo by thisisengineering, from Pexels.

p. 190: Spoon with colored sprinkles. Photo by Isabella Clifford, from Pexels.

06. MICRO-NAUTS

p. 193: Man walking on sand dune. Photo by Simon Clayton, from Pexels.

p. 194: Tree rings. Photo by Disha Sheta, from Pexels.

p. 196: Moon in blue sky. Photo by Alex Andrews, from Pexels.

p. 197: White powder on black background. Photo by Huebert World, from Pexels.

p. 198: Red Planet. Photo by SpaceX, from Pexels.

p. 200: Robotic rover. Photo by Pixabay, from Pexels.

p. 201: Hand holding clay pot. Photo by Lisa Fotios, from Pexels.

p. 202: Green tardigrade. Photo by iStock.

p. 205: View of water through cave opening. Photo by Stephan Louis, from Pexels.

p. 207: Green stalk growing out of soil. Photo by Karolina Grabowska, from Pexels.

p. 207: Silhouette of person against image of the moon. Photo by thisisengineering, from Pexels.

p. 209: Silver minerals. Photo by Pixabay, from Pexels.

p. 211: Pole with light in distance over green grass. Photo by Sunyu Kim, from Pexels.

p. 211: Silhouette of hand throwing paper airplane. Photo by Rakicevic Nenad, from Pexels.

p. 212: Phases of the moon. Photo by Alex Andrews, from Pexels.
p. 214: Silhouette of hand reaching toward crescent moon. Photo by Kaique Rocha, from Pexels.
pp. 216-17: Starry night sky. Photo by Juan González, from Pexels.
p. 218: Woman's legs walking toward a hole in the forest. Photo by Adobe.

07. ALICE IN FUTURELAND
p. 220: Woman stretching. Photo by Retha Ferguson, from Pexels.
p. 221: Woman dancing in pink haze. Photo by Retha Ferguson, from Pexels.

Index

About the Author

SPUTNIK FUTURES is a research consultancy that specializes in frontier futures: long-range intelligence that enables organizations to resonate in a world of constant and dynamic change. Sputnik has a library of original video interviews with thought leaders around the world, from Nobel Prize winners to acclaimed innovators. Sputnik partners have provided strategic foresight consultation to cross-category multinational corporations and Fortune 500 companies for more than twenty-five years and are the founders of Alice in Futureland, a platform for thought-leading content at the intersection of art, science, technology, and culture.